Centaur G-Prime Technical Description

A High-Performance Upper Stage for the NASA Space Transportation System

Centaur G-Prime Technical Description:
A High-Performance Upper Stage for the NASA Space Transportation System
(Expanded Edition, 2014)

Republication Statement

This book is a reprint of a report prepared for NASA entitled "Centaur G-Prime Technical Description: A High-Performance Upper Stage for the NASA Space Transportation System". This document is labeled Report No. GDC-SSC-83-010 (formerly CFTD-3) and was prepared for NASA by General Dynamics Convair Division. The original spiral bound document is dated January 23, 1984. This new (2014) expanded edition includes a new chapter (Section 9) of bonus material (New Content).

New Content

Section 9: A new section was added to the report. As is explained in more detail on page 9-1, Section 9 includes diagrams, photographs, newsletter articles and NASA reports. Pages 9-26 and 9-27 are derived from scans of NASA newsletters from my personal collection. Otherwise, the original pdf and jpeg files for this Report and the New Content were downloaded from the internet in May of 2014 from the NASA Technical Reports Server (NTRS) and the Defense Video and Imagery Distribution System (DVIDS). Every page from the pdf files was saved as a separate jpeg file. These jpeg and other jpeg files were further manipulated using image-editing software.

New and Modified Pages: Page (v) was modified by the addition of a listing for Section 9 in the Table of Contents. Artistic images of Shuttle-Centaur have been added to page (i), and to new page 7-16. Also, the cover is new.

Copyright

Derivative Works: The original versions of the report and the components of Section 9 are believed to be in the public domain. However, all of the pages in this reprint have been specifically enhanced to improve their legibility and black and white reproduction using image-editing software. The enhanced versions of the pictures and text contained in this publication are derivative works that are ©2014 Mooncat Collectibles.

All New Content, including New and Modified Pages, Derivative Works, and the new cover, is ©2014 Mooncat Collectibles. This publication, the Expanded Edition of *Centaur G-Prime Technical Description*, is ©2014 Mooncat Collectibles.

ISBN-13: 978-1500109479
ISBN-10: 1500109479

The cover is a montage of concept art showing Centaur deployment from the Space Shuttle superimposed on a picture of the Earth from orbit. Original versions of both images courtesy NASA; the versions used on the cover have been edited using image editing software and are derivative works that are ©2014 Mooncat Collectibles. Both images are courtesy of NASA.

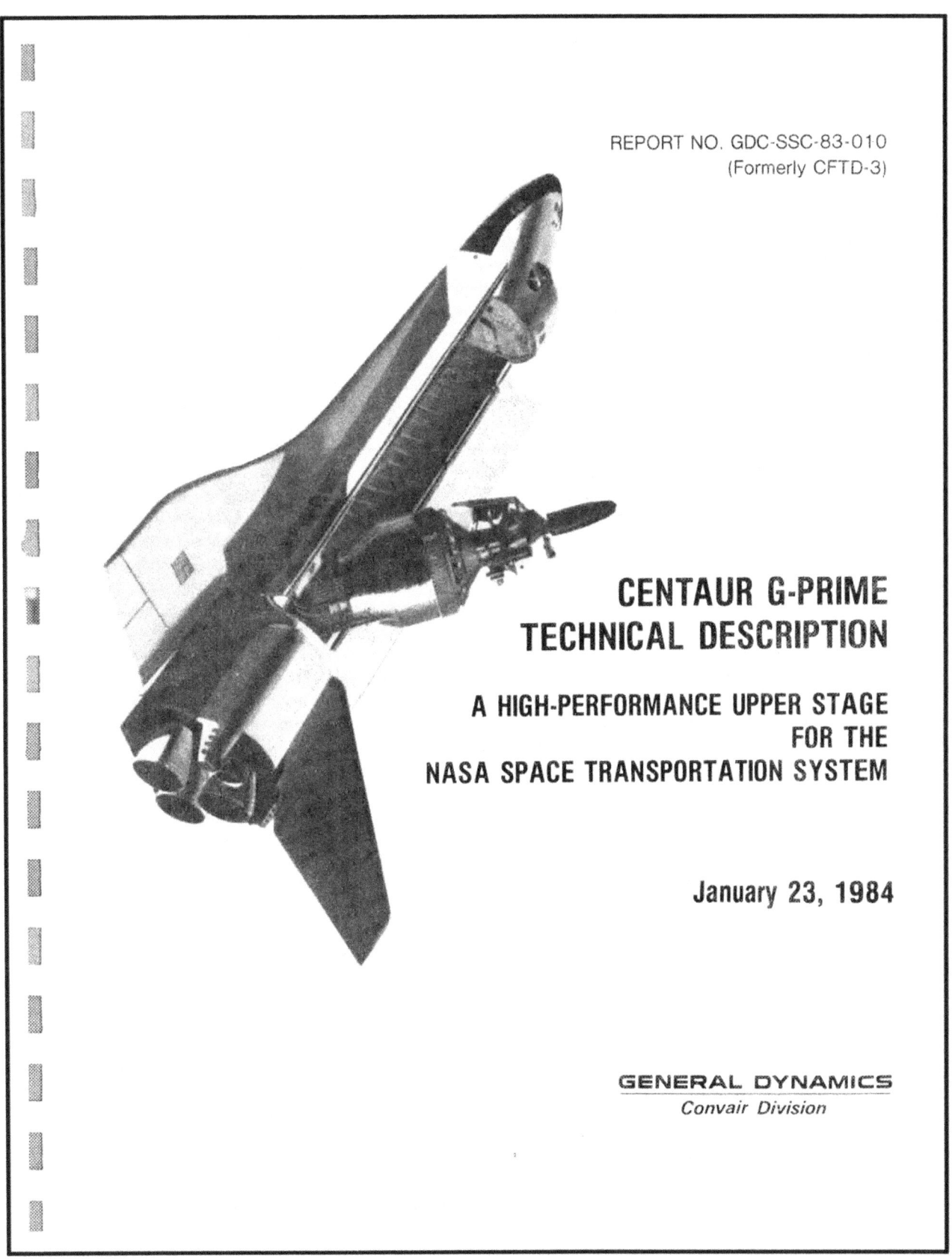

REPORT NO. GDC-SSC-83-010
(Formerly CFTD-3)

CENTAUR G-PRIME
TECHNICAL DESCRIPTION

A HIGH-PERFORMANCE UPPER STAGE
FOR THE
NASA SPACE TRANSPORTATION SYSTEM

January 23, 1984

GENERAL DYNAMICS
Convair Division

Above: Original cover for this report. Note that the original report had spiral binding.

Artist's concept of the Space Shuttle/Centaur G-prime/Galileo complex in orbit. Original image courtesy NASA; the starfield is derived from Hubble Space Telescope Image No. hs-2013-11-b and is courtesy NASA/STScI. This montage is a dertivative work that is ©2014 Mooncat Collectibles.

REPORT NO. GDC-SSC-83-010
(Formerly CFTD-3)

CENTAUR G-PRIME
TECHNICAL DESCRIPTION

A HIGH-PERFORMANCE UPPER STAGE
FOR USE IN THE
SPACE TRANSPORTATION SYSTEM

January 23, 1984

Prepared by
GENERAL DYNAMICS CONVAIR DIVISION
P.O. Box 85357
San Diego, California 92138

TABLE OF CONTENTS

Section | | | Page

1   FOREWORD   1-1

2   TECHNICAL SUMMARY   2-1

    2.1   OVERVIEW OF SHUTTLE/CENTAUR   2-1
    2.2   BASELINE MISSION AND OPTIONS   2-1
    2.3   SYSTEM DESCRIPTION   2-4

3   TECHNICAL DESCRIPTION   3-1

    3.1   GENERAL   3-1
    3.2   BASELINE MISSIONS/CONFIGURATION SELECTION   3-1
    3.2.1   One-Burn Baseline Mission   3-1
    3.2.2   Centaur G-Prime Configuration   3-2
    3.2.3   Mission Performance   3-3
    3.2.3.1   Earth-Escape Mission   3-3
    3.2.4   Flexibility for Contingencies   3-3
    3.3   CENTAUR G-PRIME CONFIGURATION   3-5
    3.3.1   Centaur System General Concept   3-5
    3.3.1.1   Centaur Vehicle   3-5
    3.3.1.2   Centaur Integrated Support System (CISS)   3-5
    3.3.2   Structural Systems   3-7
    3.3.2.1   Centaur Vehicle   3-7
    3.3.2.2   Centaur Integrated Support System (CISS)   3-23
    3.3.3   Mechanisms   3-36
    3.3.4   Fluid Systems   3-39
    3.3.4.1   Centaur Vehicle Fluid System   3-39
    3.3.4.2   CISS Fluid Systems   3-52
    3.3.5   Avionics System   3-59
    3.3.5.1   Centaur Vehicle Avionics   3-59
    3.3.5.2   CISS Avionics   3-65
    3.3.5.3   Orbiter Interfaces   3-70
    3.3.6   Software Systems   3-71
    3.3.6.1   Centaur Vehicle Software   3-72
    3.3.6.2   CISS Software   3-79
    3.3.6.3   GCS Software   3-80
    3.3.6.4   Computer-Controlled Test Equipment (CCTE) Software   3-81
    3.3.6.5   Support Software   3-81
    3.3.7   Mass Properties   3-84
    3.3.7.1   Centaur/Spacecraft Weights Summary   3-84
    3.3.7.2   Centaur Airborne Support Equipment Weights   3-84

TABLE OF CONTENTS, Contd

Section			Page
4	SYSTEMS REQUIREMENTS		4-1
	4.1	INTEGRATION REQUIREMENTS	4-1
	4.2	SYSTEM SAFETY	4-2
	4.3	RELIABILITY AND QUALITY ASSURANCE REQUIREMENTS	4-3
5	MECHANICAL FACILITIES AND GROUND SUPPORT EQUIPMENT		5-1
	5.1	GENERAL CONCEPT	5-1
	5.2	GENERAL DYNAMICS FACILITIES (SAN DIEGO)	5-2
	5.3	CAPE CANAVERAL AIR FORCE STATION (CCAFS) FACILITIES	5-2
	5.4	KENNEDY SPACE CENTER (KSC) FACILITIES	5-3
	5.5	GROUND SUPPORT EQUIPMENT	5-3
6	ELECTRICAL GROUND SYSTEMS AND EQUIPMENT		6-1
	6.1	GENERAL DYNAMICS FACILITIES	6-1
	6.1.1	Computer-Controlled Launch Set	6-1
	6.1.2	Hardware Extension Remote	6-1
	6.1.3	System Integration Laboratory	6-1
	6.1.4	Real-Time Simulation Laboratory (RTSL)	6-2
	6.1.5	Software Engineering System (SES)	6-2
	6.1.6	Computer-Controlled Test Equipment (CCTE)	6-2
	6.2	EASTERN LAUNCH SITE FACILITIES	6-3
	6.2.1	Ground Computer System	6-3
	6.2.2	Hardware Extension Remote	6-3
	6.2.3	Launch Control Functions	6-3
	6.2.4	Landline Instrumentation (LLI) System	6-3
	6.2.5	Ground Telemetry System	6-4
	6.2.6	Television (TV) Systems	6-4
	6.2.7	RF Systems	6-4
7	OPERATIONS		7-1
	7.1	GROUND OPERATIONS	7-1
	7.1.1	Normal Operations	7-1
	7.1.2	Abort Operations	7-3
	7.2	FLIGHT OPERATIONS	7-5
	7.2.1	Preflight Planning	7-5
	7.2.2	Flight Operations Support	7-6
	7.2.3	Nominal Flight	7-8
	7.2.4	Abort	7-14
	7.2.5	Postflight Evaluation	7-15

TABLE OF CONTENTS, Contd

Section			Page
8		DEVELOPMENT REQUIREMENTS	8-1
	8.1	MAJOR ANALYSIS REQUIREMENTS	8-1
	8.1.1	Structural	8-1
	8.1.2	Thermodynamics	8-3
	8.1.3	Performance	8-4
	8.1.4	Guidance	8-4
	8.1.5	Stability and Control	8-5
	8.1.6	CISS Control System Safety	8-5
	8.2	TEST PROGRAM	8-6
	8.2.1	Development Tests	8-6
	8.2.1.1	Structures	8-6
	8.2.1.2	Fluids and Mechanisms	8-6
	8.2.1.3	Avionics and Software	8-9
	8.2.1.4	Ground Avionics	8-9
	8.2.1.5	CISS/Centaur Integrated Tests	8-9
	8.2.2	Component Qualification Tests	8-9
	8.2.3	Validation Tests	8-10
9		BONUS MATERIAL (NEW CONTENT)	9-1
	9.1	INTRODUCTION	9-1
	9.2	ENHANCED FIGURES	9-2
	9.2.1	Centaur G and G-Prime	9-2
	9.2.2	CISS	9-8
	9.2.3	Avionics	9-12
	9.2.4	Flow Paths	9-14
	9.2.5	Abort Procedures	9-18
	9.3	PHOTOGRAPHS	9-20
	9.3.1	Models	9-20
	9.3.2	Rollout Ceremony	9-22
	9.3.3	Transportation (Super Guppy)	9-24
	9.4	ARTICLES	9-26
	9.4.1	Early STS-Centaur Model (picture) (1973)	9-26
	9.4.2	"Centaur Studied as Shuttle Stage" (1981)	9-27
	9.4.3	"CENTAUR FOR THE 1980s" (1981)	9-28
	9.4.4	"A High Energy Stage for the National Space Transportation System" (1984)	9-37

LIST OF FIGURES

Figure		Page
	Fluid Systems Schematic Legend	xv
2-1	Shuttle/Centaur System for Centaur G-Prime	2-1
2-2	Centaur G-Prime for Galileo and ISPM (Galileo Shown)	2-2
2-3	Crew Initiated Tasks	2-3
2-4	Centaur G-Prime Ground Operations at ELS	2-4
2-5	Shuttle/Centaur G-Prime and CISS	2-5
2-6	Centaur Integrated Support System Design Considerations	2-7
2-7	Fluid Interfaces with Orbiter	2-8
3-1	Centaur G-Prime Configuration	3-3
3-2	Centaur G-Prime Planetary Performance	3-4
3-3	Centaur Integrated Support System (CISS)	3-6
3-4	Centaur/Orbiter Fluid Interfaces	3-7
3-5	Centaur D-1A and Centaur G-Prime Tanks	3-8
3-6	Centaur G-Prime Propellant Tank Configuration	3-9
3-7	170-Inch-Diameter Centaur G-Prime LH_2 Tank	3-10
3-8	Typical Centaur G-Prime and Centaur D-1A Tank Joints	3-10
3-9	Hydrogen Tank Transition Joints	3-11
3-10	Oxygen Tank Transition Joints	3-12
3-11	Centaur G-Prime Tank Internal Installations	3-13
3-12	Centaur G-Prime Adapters	3-16
3-13	Forward Adapter	3-17
3-14	Aft Adapter	3-18
3-15	Forward Air Conditioning Duct Interface	3-18
3-16	Insulation System for Centaur G-Prime	3-20
3-17	Centaur G-Prime LH_2 Tank Sidewall Insulation Blankets	3-21
3-18	Insulation Purge and Vent Systems	3-22
3-19	Centaur Support Structure (CSS)	3-23
3-20	CSS Circular Beam Panels	3-24
3-21	CSS Circular Beam Ribs	3-25
3-22	CSS Bulkhead Installation	3-25

LIST OF FIGURES, Contd

Figure		Page
3-23	CSS Steel Interface Pins	3-26
3-24	Deployment Adapter and Engine Support Structure	3-27
3-25	Separation System	3-29
3-26	Centaur CISS Separation Guides	3-30
3-27	Galileo Hitch-up Worst Case Cabin Clearance	3-31
3-28	Deployment Adapter Shelf	3-32
3-29	CISS Helium Bottle Support Structure	3-32
3-30	CISS Air-conditioning Duct Components	3-33
3-31	CSS Adjustable Primary Longeron Fitting	3-34
3-32	CISS GN_2 Conditioning Interfaces	3-34
3-33	Active Forward Sill Latch Stop and Stretch Fittings	3-35
3-34	Active Forward Keel Latch Stop	3-36
3-35	Deployment Adapter Location System	3-37
3-36	Rotation System Functional Block Diagram	3-38
3-37	Centaur Hydraulic System	3-39
3-38	Centaur Pressurization System	3-41
3-39	Pneumatic Actuation Valve Control System (PAVCS)	3-42
3-40	Purge System	3-44
3-41	Centaur Helium Supply System	3-45
3-42	Intermediate Bulkhead Relief System	3-46
3-43	Propellant Feed and Main Engine Valve Actuation Control	3-47
3-44	Reaction Control System	3-48
3-45	Centaur Vent System	3-49
3-46	Fill/Dump Systems	3-50
3-47	Centaur/CISS Fuel and Oxidizer Disconnect Panels	3-51
3-48	5.5-inch Disconnect	3-52
3-49	CISS Vent Systems	3-53
3-50	Centaur/CISS Fill/Drain/Dump System	3-55
3-51	Propellant Tank Pressurization System	3-57
3-52	Centaur G-Prime Avionics	3-59
3-53	Failure Tolerant Timed Arming Sequence	3-62
3-54	Baseline Centaur G-Prime Command Link Paths	3-63

LIST OF FIGURES, Contd

Figure		Page
3-55	Centaur G-Prime Instrumentation System	3-64
3-56	Centaur G-Prime Telemetry System	3-65
3-57	Centaur G-Prime Electrical Power System	3-66
3-58	CISS Avionics	3-67
3-59	Five-string Voter Ensures Two-Failure Tolerance	3-68
3-60	CISS Instrumentation System for Centaur G-Prime	3-69
3-61	Shuttle/Centaur Software System	3-72
3-62	Centaur DCU Software Checkout	3-78
5-1	Prelaunch Operations	5-1
5-2	Centaur G-Prime Abort Flow Path	5-2
6-1	Ground Systems Overview	6-2
7-1	Centaur G-Prime CISS Prelaunch Flowpath	7-2
7-2	Centaur G-Prime Processing Flow at CCAFS Facilities	7-2
7-3	Centaur G-Prime Flow Path at KSC	7-3
7-4	Post Flight CISS Flowpath	7-4
7-5	Centaur G-Prime Abort Flow Path	7-4
7-6	CPOCC Support for NASA Mission Flight Operations	7-7
7-7	Orbiter Crew Control of Critical Functions	7-8
7-8	Nominal CCE Ascent Operations	7-9
7-9	Rev 4 Deployment Timeline	7-9
7-10	Nominal/Centaur Postdeployment Operations	7-10
7-11	Centaur Can Dump Propellants Safely in All Abort Modes	7-14
7-12	Zero-G Propellant Dump	7-16
8-1	Centaur G-Prime Forward Adapter Finite Element Model	8-1
8-2	Centaur G-Prime and CISS Finite Element Model	8-2
8-3	Centaur G-Prime Test Program Flow Diagram	8-16
8-4	CISS/Centaur Test Flow	8-17
FO-1	Combined Fluid Systems Schematic	FO-1

LIST OF TABLES

Table		Page
2-1	Centaur G-Prime Representative Mission Events Before MES	2-2
3-1	Mission Requirements	3-2
3-2	Centaur G-Prime Mission Weight Summaries	3-4
3-3	Preflight and Flight Software Modules	3-74
3-4	Estimated Centaur Vehicle DCU Memory Requirements	3-75
3-5	Control Unit Functional Control Modules	3-80
3-6	Centaur Airborne Support Equipment Sum	3-84
7-1	Centaur G-Prime Flight Operations Requirements	7-11
8-1	Typical FOMs (msec) for Galileo and ISPM Missions	8-5
8-2	Test Program Summary	8-7

ACRONYMS AND ABBREVIATIONS

A/C	Atlas/Centaur
AFD	Aft Flight Deck
AFETR	Air Force Eastern Test Range
AFO	Abort From Orbit
AOA	Abort Once Around
ASE	Airborne Support Equipment
ATO	Abort to Orbit
CAD	Computer-Aided Design
CASE	Centaur Airborne Support Equipment
CCA	Centaur/CISS Assembly
CCAFS	Cape Canaveral Air Force Station
CCC	Command/Control/Communication
CCE	Centaur Cargo Element
CCLS	Computer Controlled Launch Set
CCT	Centaur CISS Transporter
CCTE	Computer Controlled Test Equipment
CCU	Central Control Unit
CCVAPS	Computer Controlled Vent and Pressurization System
CDR	Critical Design Review
CDU	Control Distribution Unit
CG	Center of Gravity
CIL	Critical Items List
CISS	Centaur Integrated Support System
CITE	Cargo Integration Test Equipment
CPOCC	Centaur Payload Operations Control Center
CPU	Central Processing Unit
CRT	Cathode Ray Tube
CRTP	Centaur Transport Pallet
CSS	Centaur Support Structure
CSTP	CISS Transport Pallet
CU	Control Unit
CWEA	Caution and Warning Electronic Assembly
C&W	Caution and Warning
Cx	Complex
C_3	Orbital Energy
DA	Deployment Adapter
DAC	Digital Analog Converter
DCU	Digital Computer Unit
DDR	Digital Derived Rate
DDT&E	Design Development Test & Evaluation
DET	Design Evaluation Test
DMA	Direct Memory Access
DoD	Department of Defense
DOF	Degrees of Freedom

ACRONYMS AND ABBREVIATIONS, Contd

DPT	Design Proof Test
DSN	Deep Space Network
DSO	Discrete Signal Output
DUFTAS	Dual-Failure-Tolerant Arm/Safe Sequencer
ECS	Environment Control System
EDU	Electrical Distribution Unit
EFH	Extra Full Hard
EGA	Earth Gravity Assist
ELS	Eastern Launch Site
EMC	Electromagnetic Compatibility
EMI	Electromagnetic Interference
ESMC	Eastern Space and Missile Center
ET	External Tank
ETR	Eastern Test Range
EVA	Extravehicular Activity
FAST	Flight Analogous Software Test
FCR	Flight Control Room
FM	Frequency Modulated
FMEA	Failure Modes and Effects Analysis
FOM	Figure of Merit
FPR	Flight Performance Reserve
FRD	Functional Requirements Document
FRR	Functional Requirements Review
FSE	Fixed Support Equipment
FTA	Fault Tree Analysis
GCS	Ground Computer System
GDC	General Dynamics Convair
G&N	Guidance & Navigation
GEO	Geostationary Earth Orbit
GET	Ground Elapsed Time
GFP	Government-Furnished Property
GHe	Gaseous Helium
GH_2	Gaseous Hydrogen
GLL	Galileo
GN_2	Gaseous Nitrogen
GN&C	Guidance, Navigation, and Control
GO_2	Gaseous Oxygen
GPC	General Purpose Computer
GSE	Ground Support Equipment
GTS	Ground Telemetry SYstem
HCS	Helium Control Skid
HER	Hardware Extension Remote
HRM	Hydraulic Recirculation Motors
IAR	Integration Analysis Review

ACRONYMS AND ABBREVIATIONS, Contd

IAT	Initial Acceptance Test
ICD	Interface Control Document
ICS	Interpretive Computer Simulator
IDR	Intermediate Design Review
IMG	Inertial Measurement Group
I/O	Input/Output
IRGU	Integrating Rate Gyro Unit
IRU	Inertial Reference Unit
ISPM	International Solar Polar Mission
ITP	Integrated Test Plan
IUS	Inertial Upper Stage
JPL	Jet Propulsion Laboratory
JPM	Jet Pulse Mixer
JSC	Johnson Space Center
kbps	kilo (thousands) bits per second
KSC	Kennedy Space Center
kW	Kilowatt
lbf	Pounds Force
LCC	Launch Control Center
LDIE	Local Digital Interface Electronics
LEO	Low Earth Orbit
LeRC	Lewis Research Center
LH_2	Liquid Hydrogen
LHCS	Liquid Hydrogen Control Skid
LMSC	Lockheed Missile & Space Company
LN_2	Liquid Nitrogen
LOCS	Liquid Oxygen Control Skid
LO_2	Liquid Oxygen
LSI	Launch System Integration
LSSP	Launch Site Support Plan
LVC	Launch Vehicle Contingency Propellants
LVMP	Launch Vehicle Mission Peculiar
MCC	Mission Control center
MCCC	Mission Control and Computer Center (JPL)
MDAC	McDonnell Douglas Astronautics Company
MDM	Multiplexer Demultiplexer Multiplex
MECO	Main Engine Cutoff
MES	Main Engine Start
MET	Mission Elapsed Time
MIL	GSFC Spaceflight Tracking and Data Network Station (KSC)
MLP	Mobile Launch Platform
MMSE	Multi-Use Mission Support Equipment
MPSR	Multi-Purpose Support Room
MSE	Mobile Support Equipment
MSS	Mission Specialist Station
MUX	Multiplexer

ACRONYMS AND ABBREVIATIONS, Contd

NEI	Non-explosive Initiators
NGT	NASA Ground Terminal
NPSH	Net Positive Suction Head
NSI	NASA Standard Initiator
N_2H_4	Hydrazine
OMS	Orbital Maneuvering System
OPF	Orbiter Processing Facility
PAST	Preflight Analogous Software Test
PCM	Pulse Code Modulation
PCOS	Power Changeover Switching
PCR	Payload Changeout Room
PDI	Payload Data Interleaver
PDR	Preliminary Design Review
PE	Propellant Excess
PFAST	Preflight Analogous Software Test
PGHM	Payload Ground Handling Mechanism
PI	Payload Interrogator
PICU	Pyro Initiator Control Unit
PIP	Payload Integration Plan
P/L	Payload
PLIS	Propellant Loading Indicator System
PLIU	Propellant Level Indicating Unit
PLS	Propellant Loading System
PM	Propellant Margin
POCC	Payload Operational Control Center
PROM	Programmable Read-Only Memory
psid	Pounds Per Square Inch - Differential
psig	Pounds Per Square Inch - Gage
PSP	Payload Signal Processor
PTC	Passive Thermal Control
PU	Propellant Utilization
PWA	Pratt & Whitney Aircraft
RAM	Random Access Memory
RCS	Reaction Control System
RF	Radio Frequency
RIC	Remote Interface Controllers
RMS	Remote Manipulator System
RMU	Remote Multiplexer Unit
RSS	Rotating Service Structure
RTLS	Return to Launch Site
RTSL	Real-Time Simulation Laboratory
SAEF	Spacecraft Assembly & Encapsulation Facility
S/C	Spacecraft
SC	Signal Conditioner
SCU	Sequence Control Unit
SEF	Software Engineering Facility
SES	Software Engineering System

ACRONYMS AND ABBREVIATIONS, Contd

SEU	System Electronics Unit
SGLS	Space Ground Link System
SID	System Interface Document
SIL	System Integration Laboratory
SIR	System Interface Requirements
SIRD	Support Instrumentation Requirements Document
SIU	Servo Inverter Unit
SLF	Shuttle Landing Facility
SMCH	Standard Mix Cargo Harness
SPCU	Standby Pneumatic Control Unit
SPAPL	Space Programs Approved Parts List
SPIF	Shuttle Payload Integration Facility
SRB	Solid Rocket Booster
SRM	Solid Rocket Motor
SSP	Standard Switch Panel
Sta	Station
STAR	Shuttle Turnaround Assessment Report
STE	Special Test Equipment
STDN	Space Tracking & Data Network
STS	Space Transportation System
S/W	Software
TAL	Trans-Atlantic Landing
TALA	Trans-Atlantic Landing Avoidance
TBD	To Be Determined
T/C	Titan/Centaur
TDRSS	Tracking and Data Relay Satellite System
TEMPEST	Secured Communications Code Word
TIU	Telemetry Interface Unit
TLM	Telemetry
TTF	Test and Transport Fixture
TVS	Thermodynamic Vent System
UDU	Uplink/Downlink Unit
UPS	Uninterruptible Power Source
VAB	Vehicle Assembly Building
VAFB	Vandenberg AFB
VOIR	Venus Orbiting Imaging Radar
VPF	Vertical Processing Facility
VPHD	Vertical Processing Handling Device
WSGT	White Sands Ground Terminal
XMT	Transmitter
ZOE	Zone of Exclusion

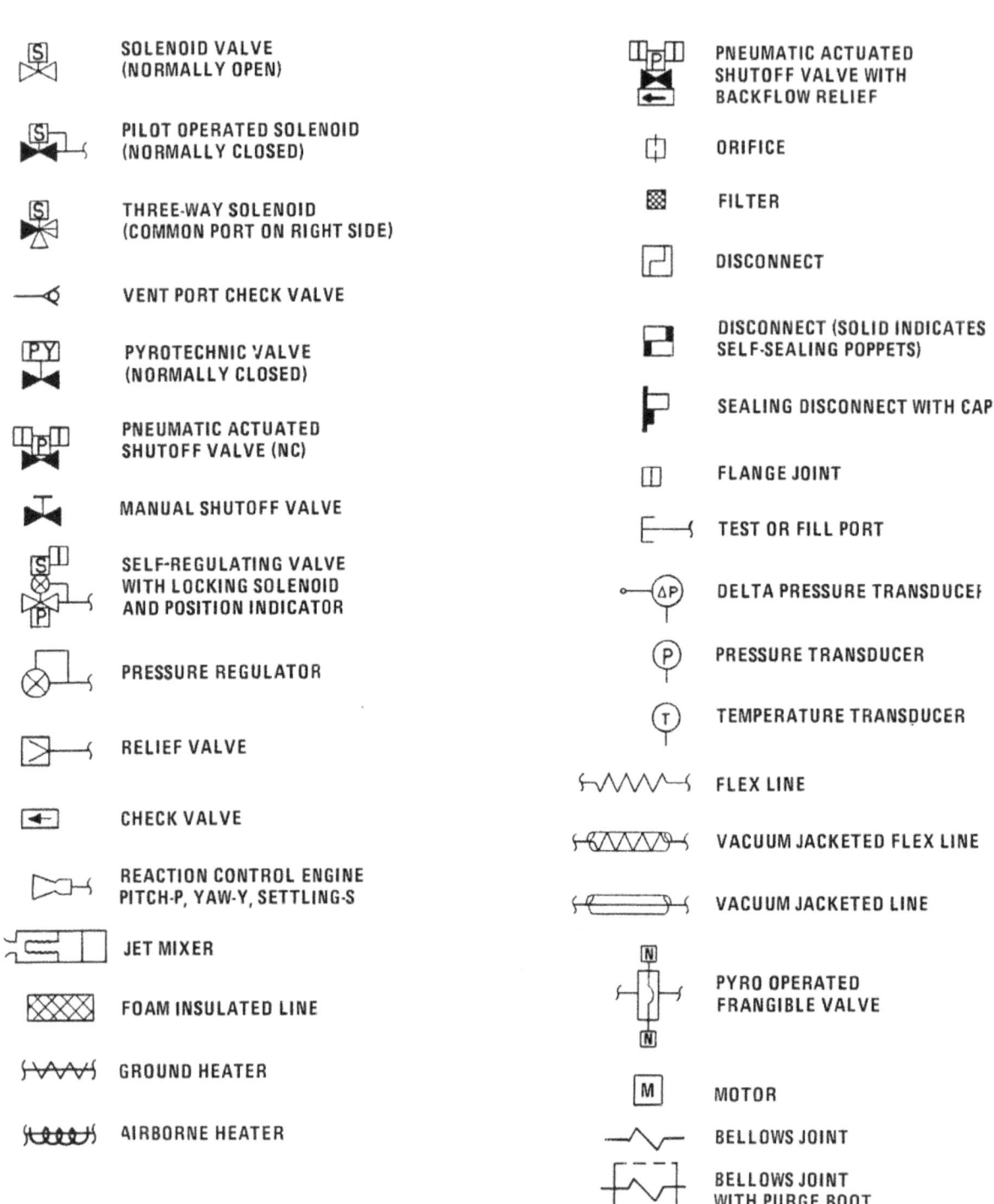

SOLENOID VALVE (NORMALLY OPEN)	PNEUMATIC ACTUATED SHUTOFF VALVE WITH BACKFLOW RELIEF
PILOT OPERATED SOLENOID (NORMALLY CLOSED)	ORIFICE
THREE-WAY SOLENOID (COMMON PORT ON RIGHT SIDE)	FILTER
VENT PORT CHECK VALVE	DISCONNECT
PYROTECHNIC VALVE (NORMALLY CLOSED)	DISCONNECT (SOLID INDICATES SELF-SEALING POPPETS)
PNEUMATIC ACTUATED SHUTOFF VALVE (NC)	SEALING DISCONNECT WITH CAP
MANUAL SHUTOFF VALVE	FLANGE JOINT
SELF-REGULATING VALVE WITH LOCKING SOLENOID AND POSITION INDICATOR	TEST OR FILL PORT
PRESSURE REGULATOR	DELTA PRESSURE TRANSDUCER
RELIEF VALVE	PRESSURE TRANSDUCER
CHECK VALVE	TEMPERATURE TRANSDUCER
REACTION CONTROL ENGINE PITCH-P, YAW-Y, SETTLING-S	FLEX LINE
JET MIXER	VACUUM JACKETED FLEX LINE
FOAM INSULATED LINE	VACUUM JACKETED LINE
GROUND HEATER	PYRO OPERATED FRANGIBLE VALVE
AIRBORNE HEATER	MOTOR
	BELLOWS JOINT
	BELLOWS JOINT WITH PURGE BOOT

267.509-1

Fluid Systems Schematic Legend

SECTION 1

FOREWORD

Integration of the Centaur vehicle into the Space Shuttle offers a significant increase in the performance capability of the Space Transportation System. During the last several years, substantial contractor and NASA activity (General Dynamics Convair Division and Rockwell; LeRC, JSC, and KSC) led to NASA's conclusion that the Centaur vehicle can be integrated into the Shuttle. This NASA Centaur vehicle has been designated Centaur G-prime.

The vehicle described in this document can perform the Galileo and Solar Polar missions for NASA. Centaur G-Prime takes advantage of the Shuttle's 15-foot-diameter payload bay and the 65,000-pound lift capability of the Shuttle system.

The configuration is derived from the flight-proven Centaur D-1A and Centaur D-1T vehicles. Basically, the 120-inch diameter LO_2 tank is lengthened and a conical section is added to transition to the 170-inch-diameter hydrogen tank. The advantage of this configuration is it provides increased propellant capacity while accommodating longer payload lengths than a comparable 120-inch-diameter vehicle.

The following chapters present Shuttle/Centaur mission considerations, describe the Shuttle/Centaur airborne and ground hardware/software configurations and technical approaches, and outline the integrated ground and flight operations.

Shuttle/Centaur configurations and approaches discussed in this document are based on current development and integration planning. The major ground rules and assumptions factored into these configurations include:

1. Previous safety review results remain applicable.

2. Impact on Shuttle hardware/software and facilities will be minimized.

3. Doors-closed abort duration will be no greater than 6.5 hours.

4. Centaur/spacecraft will be installed in the Orbiter payload bay via the payload ground handling mechanism (PGHM).

5. JSC 07700, Volume XIV, Revision G, dated 26 September 1980, provides applicable environmental design requirements; e.g., acoustic noise levels, etc.

SECTION 2

TECHNICAL SUMMARY

2.1 OVERVIEW OF SHUTTLE/CENTAUR

Integration of the flight-proven Centaur into the Space Shuttle is being accomplished with minimum modification to Centaur or the Shuttle by using the Centaur Integrated Support System (CISS), as illustrated in Figure 2-1. The requirements of the NASA Galileo (GLL) mission and International Solar Polar mission (ISPM) (Figure 2-2) are met by the Centaur G-prime configuration. Basic Shuttle integration and safety considerations have necessitated some additional changes to the present Centaur D-1A configuration.

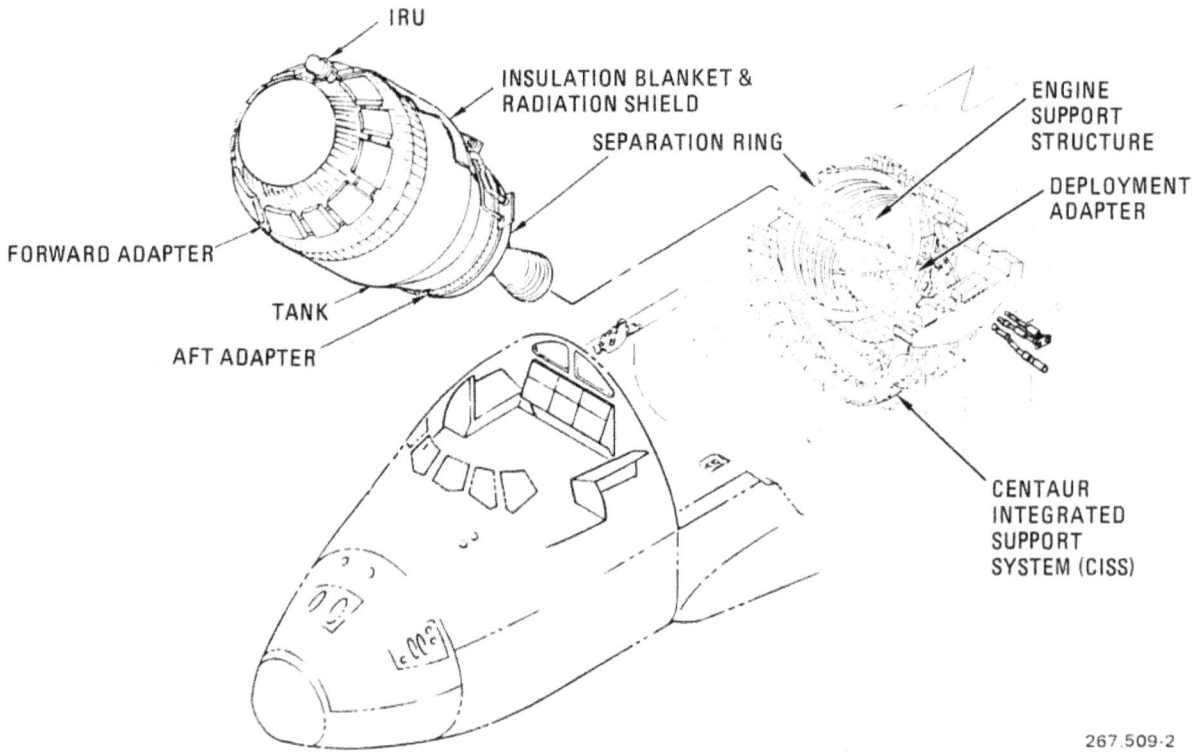

267.509-2

Figure 2-1. Shuttle/Centaur System for Centaur G-Prime

2.2 BASELINE MISSION AND OPTIONS

The baseline mission incorporates a single burn of the Centaur main engines to inject the spacecraft into an earth-escape trajectory. This burn occurs no earlier than 45 minutes after separation from the Orbiter, which has been inserted into a 130-n.mi. circular parking orbit. Table 2-1 summarizes major events occurring before the Centaur burn. This mission profile has been developed for the NASA Galileo and ISPM missions.

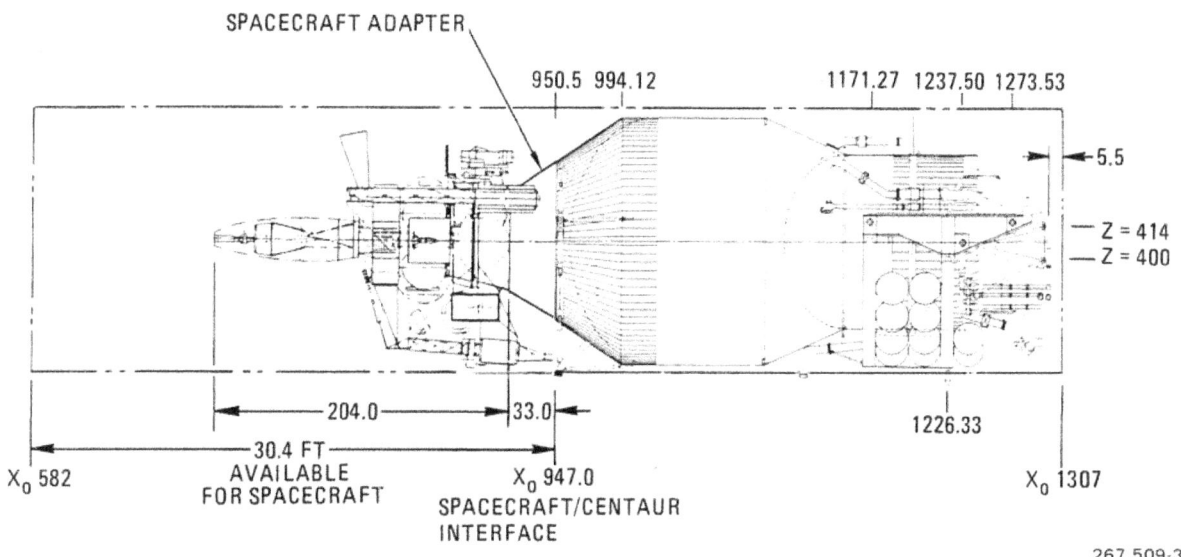

Figure 2-2. Centaur G-Prime for Galileo and ISPM (Galileo Shown)

Table 2-1. Centaur G-Prime Representative Mission Events Before MES

Event	Time
Ascent	Time from Liftoff
Shuttle liftoff	0
Jettison SRBs	2 min. 4 sec.
Shuttle MECO	8 min. 11 sec.
ET Separation	8 min. 16 sec.
OMS-1	8 min. 54 sec.
OMS-2	40 min. 16 sec.
Open Payload Bay Doors	1 hr. 3 min.
Initiate CCE On-Orbit Mode	1 hr. 13 min.
Deployment	Time to Centaur MES
Initiate Centaur Checkout	5 hr. 8 min.
Go/no-go to Deploy	1 hr. 19 min.
Rotate Deployment Adapter	1 hr. 17 min.
Commit Centaur/Spacecraft	1 hr. 6 min.
Go/no-go to Separate	1 hr. 3 min.
Separation from Orbiter	50 min.
Parking Orbit Coast	Time to Centaur MES
Activate Centaur RCS	45 min.
Perform Spacecraft Pointing Maneuvers	45 min.
Orient to Centaur burn attitude	5 min.
MES	0

Centaur predeployment events are controlled automatically from the CISS, with crew initiation interspersed as necessary to initiate on-orbit deployment or safing functions (Figure 2-3).

Mission flexibility is readily accommodated by present Centaur D-1A software logic and structuring with only minor changes anticipated to meet mission-peculiar requirements. The approach includes preflight ground initialization of the Centaur general-purpose computer and prelaunch calibration of the inertial measurement unit. Both then function continuously until the end of the mission.

The overall ground operations flow required to check out the Centaur/CISS thoroughly, mate the spacecraft, and assemble and launch them in the Space Shuttle is illustrated in Figure 2-4.

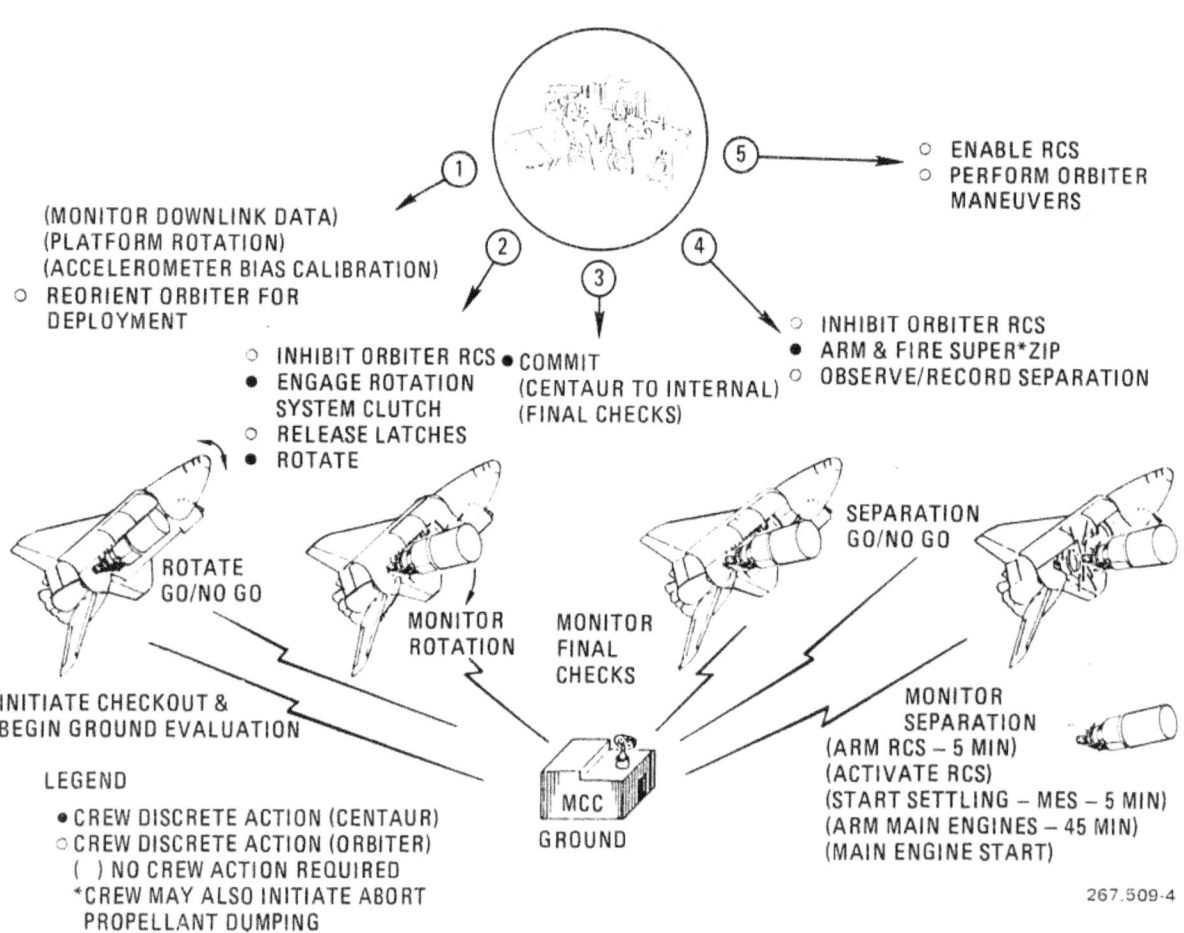

Figure 2-3. Crew Initiated Tasks

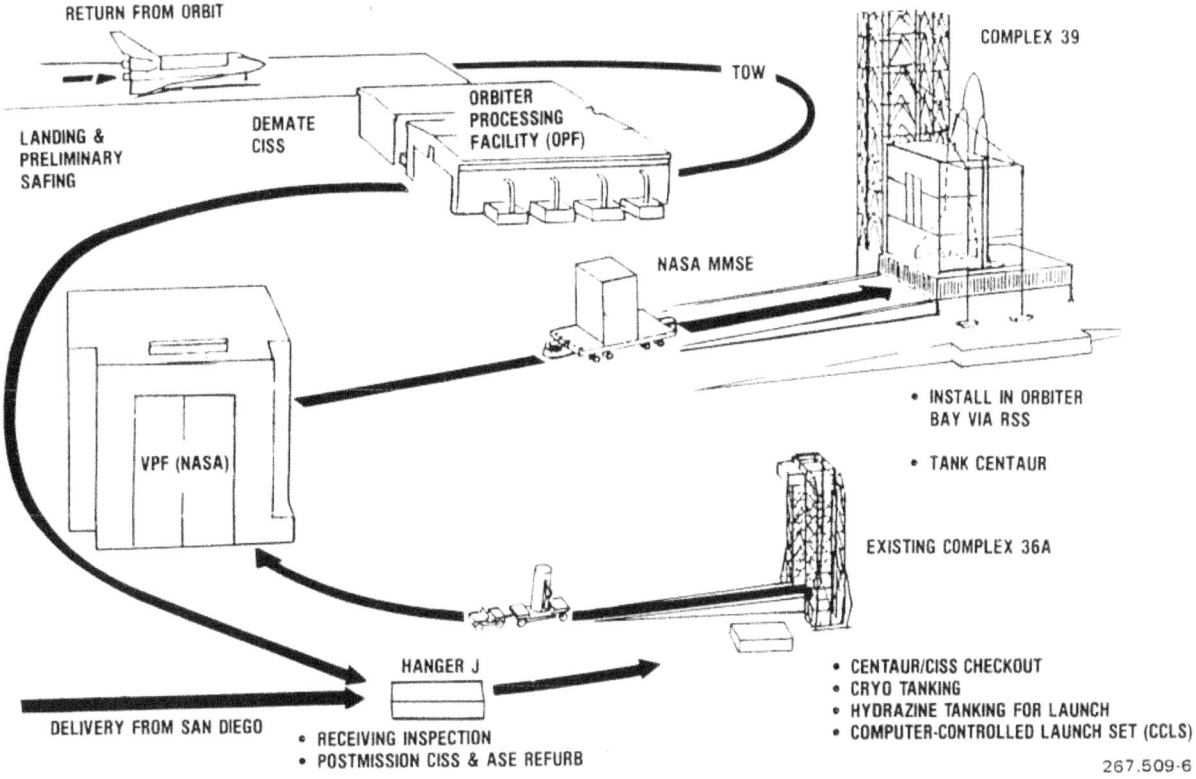

RETURN FROM ORBIT

LANDING &
PRELIMINARY
SAFING

DEMATE
CISS

ORBITER
PROCESSING
FACILITY (OPF)

TOW

COMPLEX 39

NASA MMSE

VPF (NASA)

• INSTALL IN ORBITER
  BAY VIA RSS

• TANK CENTAUR

EXISTING COMPLEX 36A

HANGER J

DELIVERY FROM SAN DIEGO

• RECEIVING INSPECTION
• POSTMISSION CISS & ASE REFURB

• CENTAUR/CISS CHECKOUT
• CRYO TANKING
• HYDRAZINE TANKING FOR LAUNCH
• COMPUTER-CONTROLLED LAUNCH SET (CCLS)

267.509-6

Figure 2-4. Centaur G-Prime Ground Operations at ELS

2.3 SYSTEM DESCRIPTION

The overall Centaur G-prime vehicle is 29.1 feet long with a maximum LH_2 tank diameter of 170 inches (Figure 2-5). With the exception of the deployable antennae, external protuberances (rings, stringers, insulation, harnessing, and fluid lines) do not violate the 180-inch payload envelope under any identified dynamic or thermal environmental conditions when Centaur G-prime is supported within the Orbiter.

The Centaur vehicle includes a forward adapter attached to the LH_2 tank. This adapter supports Centaur avionics, provides a mounting interface for the spacecraft, and reacts the loads between Centaur and Orbiter forward attachments. An aft adapter and separation ring attached to the aft end of Centaur distribute acceptable circumferential line loads into the tank and provide pyrotechnic-actuated separation for Centaur deployment from the Orbiter.

The pressure-stabilized Centaur G-prime propellant tank will be constructed in a manner very similar to that of the present Centaur D-1A, although there are geometry differences.

Construction techniques for welding high-strength stainless steel have proven highly reliable for Centuar D-1A for more than fifteen years. The double-wall intermediate bulkhead is identical on all Atlas/Centaur and Shuttle/Centaur tanks, with but minor differences at the cylindrical ring joints.

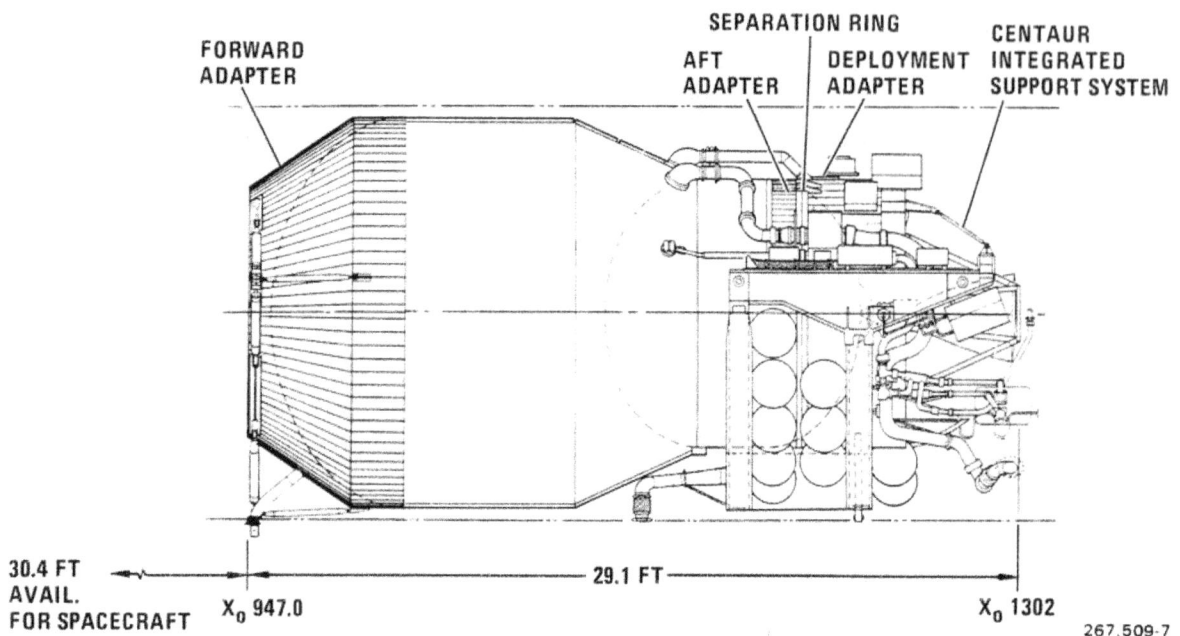

Figure 2-5. Shuttle/Centaur G-Prime and CISS

Similarly, the aft bulkhead is identical on all Centaur tanks, thereby retaining GDC's test and flight experience and minimizing modifications to flight-proven fluid systems.

The forward and aft adapters are similar in function to structures flown on Centaur D-1A and Centaur D-1T. The cylindrical section of the forward adapter and the aft adapter will be constructed of graphite epoxy composite material to enhance performance.

Centaur's insulation system consists of two polyimide, fire-resistant foam blankets enclosed by a multilayer radiation shield/helium containment membrane over the entire LH_2 tank. The LH_2 tank insulation blanket is purged with helium before launch.

The Centaur LO_2 tank aft bulkhead supports the two RL10A3-3A engines and the associated LO_2 and LH_2 propellant supply systems. The RL10 engines will be operated at the presently qualified nominal mixture ratio of 5:1 and a thrust of 16,500 lb.

The aft bulkhead also supports the hydrazine monopropellant reaction control system and the engine-mounted hydraulic system that supplies power to gimbal the main engines to effect flight control. The bulkhead also supports pneumatic storage and supply systems and the tank vent systems used to control tank pressurization.

Propellants are supplied to the main engines with specified net positive suction head (NPSH) provided by tank pressurization. Tank pressure is controlled by a computer-controlled vent and pressurization system (CCVAPS), which injects helium into the tanks before engine start in response to sensed tank pressures. After engine start, the LH_2 tank is pressurized with GH_2, which is bled from the engines. The LO_2 tank is pressurized with helium. The systems are essentially identical for Centaur D-1A and Centaur G-prime vehicles, with the minor exception of line routing, geometry differences, and redundancy (for safety).

The Centaur avionics system consists of a 16K core memory digital computer unit, a four-gimballed-platform inertial measurement group, sequence control unit, servo inverter unit, two remote multiplexing units, two signal conditioners, pyrotechnic initiator control units, associated instrumentation system, propellant utilization and level-sensing system, telemetry system, and an electrical power system and batteries. These systems/units operate together to control all vehicle functions.

Shuttle integration and safety requirements have necessitated a few minor component changes to the Centaur D-1A avionics system. Where practical, commonality among Centaur D-1A and Centaur G-prime configurations is maintained. Safety requirements dictate addition of a dual-failure-tolerant arm/safe sequencer (DUFTAS), which pecludes premature arming of critical Centaur and spacecraft functions. Mission peculiar avionics changes to meet future requirements can be accommmodated, such as an optional star scan capability.

The Centaur vehicle is supported and serviced while within the Orbiter payload bay via the Centaur Integrated Support System (CISS), as indicated in Figure 2-6. The CISS includes a deployment adapter that supports the Centaur vehicle at the separation ring. The adapter also supports the Centaur main engines via a truss structure on the aft end and supports various avionic, mechanical, and fluid subsystems and components. The deployment adapter attaches to the Centaur support structure (CSS) via two rotation trunnions. The support structure transfers loads between Centaur and the Orbiter through a five-point support system. It also supports other avionic, mechanical, and fluid subsystems, including systems serving as interfaces between the Orbiter and Centaur.

CASE functional systems include mechanical systems to:

● Rotate the deployment adapter to a separation attitude of 45 degrees.

● Retain the Centaur & CISS within the Orbiter payload bay.

CISS also incorporates fluid systems to:

● Fill and drain the LO_2 and LH_2 propellants on the ground.

● Dump LH_2 and LO_2 propellants in flight.

● Provide helium to Centaur for purging, pneumatic control system pressure, and tank pressurization.

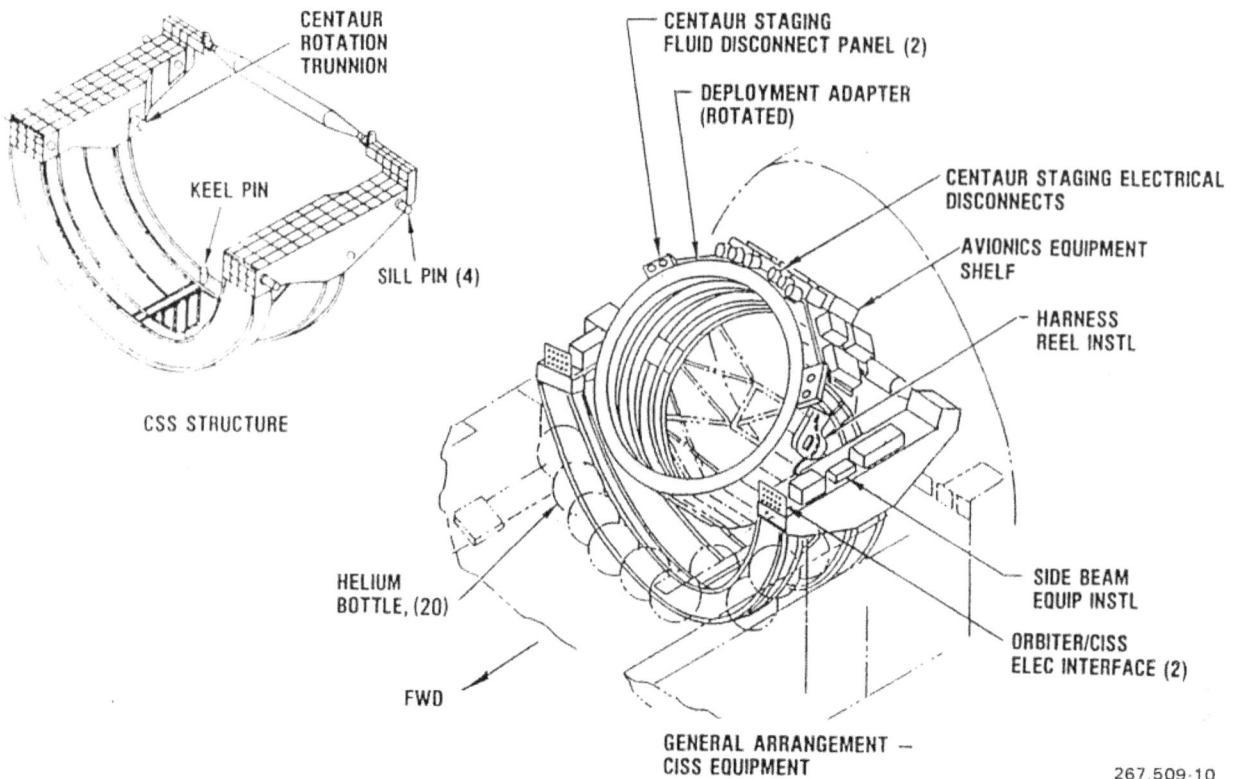

Figure 2-6.   Centaur Integrated Support System Design Considerations

- Vent LO$_2$ and LH$_2$ tanks on the ground and in flight.

- Distribute prelaunch GN$_2$ gas conditioning on the ground.

CISS avionics systems include:

- Five control units and two control distribution units that provide two-failure-tolerant control of all CISS/Centaur subsystems before deployment.

- A redundant power subsystem to augment Orbiter-supplied electrical power.

- An instrumentation system to distribute data.

The baseline CISS avionics configuration is shown in Section 3, Figure 3-62.

Shuttle Integration Considerations

Shuttle Centaur has been subjected to Phase 0 and I safety reviews by NASA/JSC and NASA/KSC. It was concluded Centaur can be integrated safely into the Space Shuttle. Phase II and III safety reviews will occur during the Centaur integration effort.

Major Centaur-to-Orbiter fluid interfaces are shown in Figure 2-7.

During Orbiter-initiated aborts, the Centaur propellants will be dumped. This provides several significant advantages:

- Return (landing) cargo weight is reduced by nearly 45,000 pounds.

- Abort to orbit capability occurs earlier in the Orbiter trajectory if propellants are dumped. This ensures an overlap between the RTLS and ATO abort modes, thus avoiding a trans-Atlantic abort.

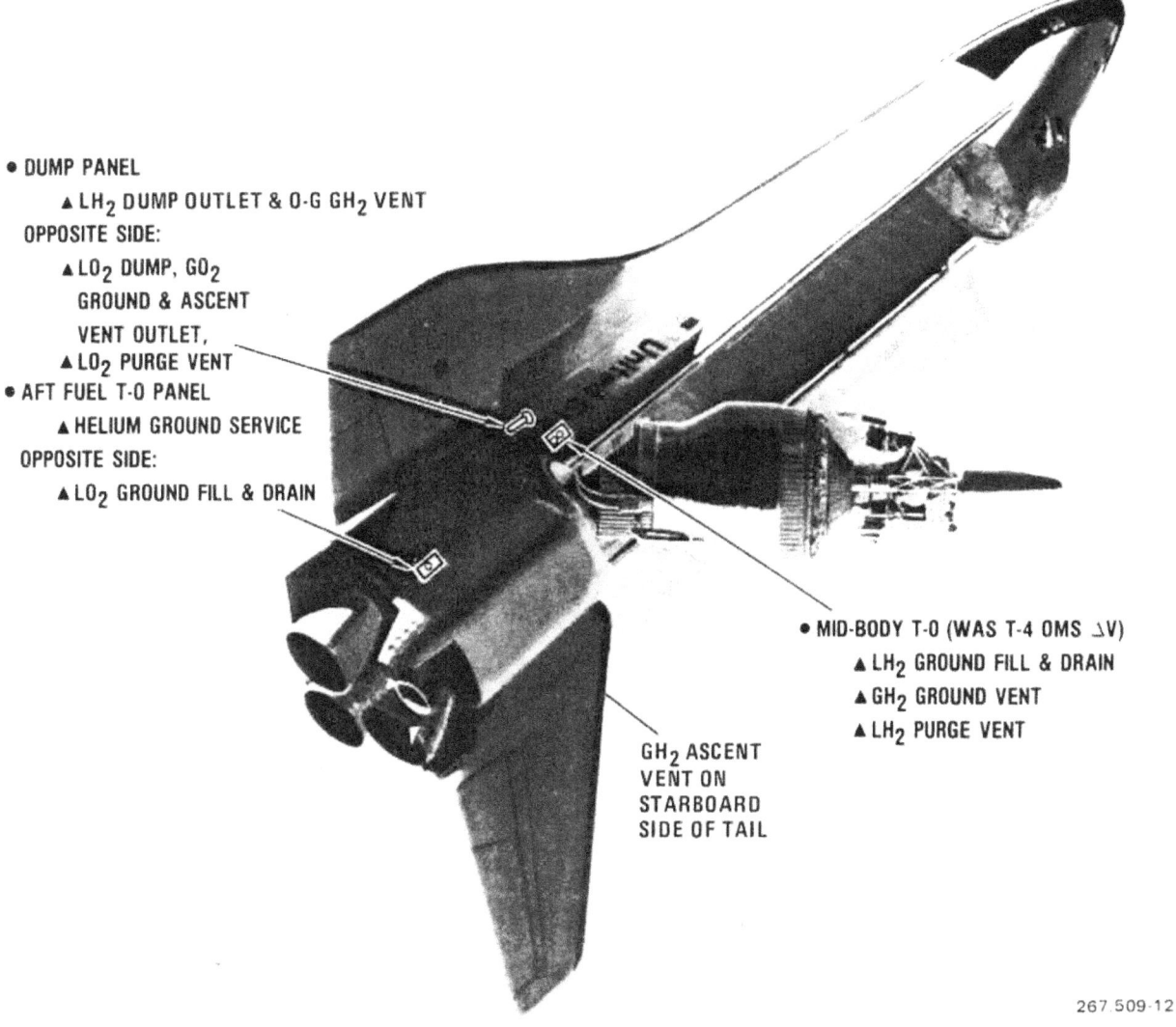

Figure 2-7. Fluid Interfaces with Orbiter

SECTION 3

TECHNICAL DESCRIPTION

3.1 GENERAL

This chapter describes technical features of the Centaur G-prime General
Dynamics is developing for NASA. The baseline Centaur G-prime and
mission-peculiar modifications required for various mission options are
covered.

Design, development, test, and evaluation of Centaur G-prime described in this
chapter include:

- Mission requirements development.

- Centaur flight vehicle design, development, analysis, software
  development, and mission peculiars.

- Centaur integrated support system design, development, analysis, software
  development, and mission peculiars.

- Ground support equipment and facilities design, development, analysis, and
  software development.

- Ground and flight operations development and planning.

- Hardware and software development and qualification testing.

3.2 BASELINE MISSIONS/CONFIGURATION SELECTION

The Centaur G-prime baseline mission is a one-burn Galileo and ISPM mission.

3.2.1 ONE-BURN BASELINE MISSION. The baseline reference NASA mission adopted
for engineering design and analysis of Centaur G-prime is the planetary 1986
Galileo launch, achieved with a single Centaur burn from a 130-n.mi. parking
orbit provided by the STS. This burn places the spacecraft in a heliocentric
orbit. The Galileo spacecraft encounters the planet Jupiter two years after
launch. Requirements for this NASA baseline mission and the similar ISPM
mission are summarized in Table 3-1.

Table 3-1. Mission Requirements

| | NASA Application | |
Item	Galileo	ISPM
● Parking orbit:		
130-n.mi. circular	X	X
28.5-deg inclination	X	X
(due-east launch)		
● Earth-escape (hyperbolic)		
C_3 80 km^2/sec^2	X	
C_3 133 km^2/sec^2		X
● Nominal Centaur deployment		
Fourth park orbit revolution	X	X
● Backup deployment opportunities:		
1 & 2 revolutions past nominal	X	X
● Spacecraft weight:		
5302 lb	X	
809 lb		X
● Spacecraft length:		
Up to 30.4 ft	X	X

3.2.2 CENTAUR G-PRIME CONFIGURATION. An assessment of the NASA planetary mission requirements has led to selection of a Centaur configuration to satisfy the NASA one-burn mission and be readily adaptable to the two-burn optional missions with minimal mission-peculiar differences. This Centaur configuration is characterized by:

● Approximately 29.1-foot overall length.

● Approximately 46,000 lb total propellant capacity.

● Standard RL-10 engines (nominal mixture ratio of 5:1 and specific impulse of 446.4 seconds).

● Standard ellipsoidal bulkheads and 24-degree transition angle from the LH_2 tank 170-inch diameter to the LO_2 tank 120-inch diameter.

The selected configuration (Figure 3-1) accommodates a 30.4-foot payload length in the payload bay. Since performance with the well-proven Centaur propulsion system is adequate for the baseline missions, the standard-length engine bells will be used.

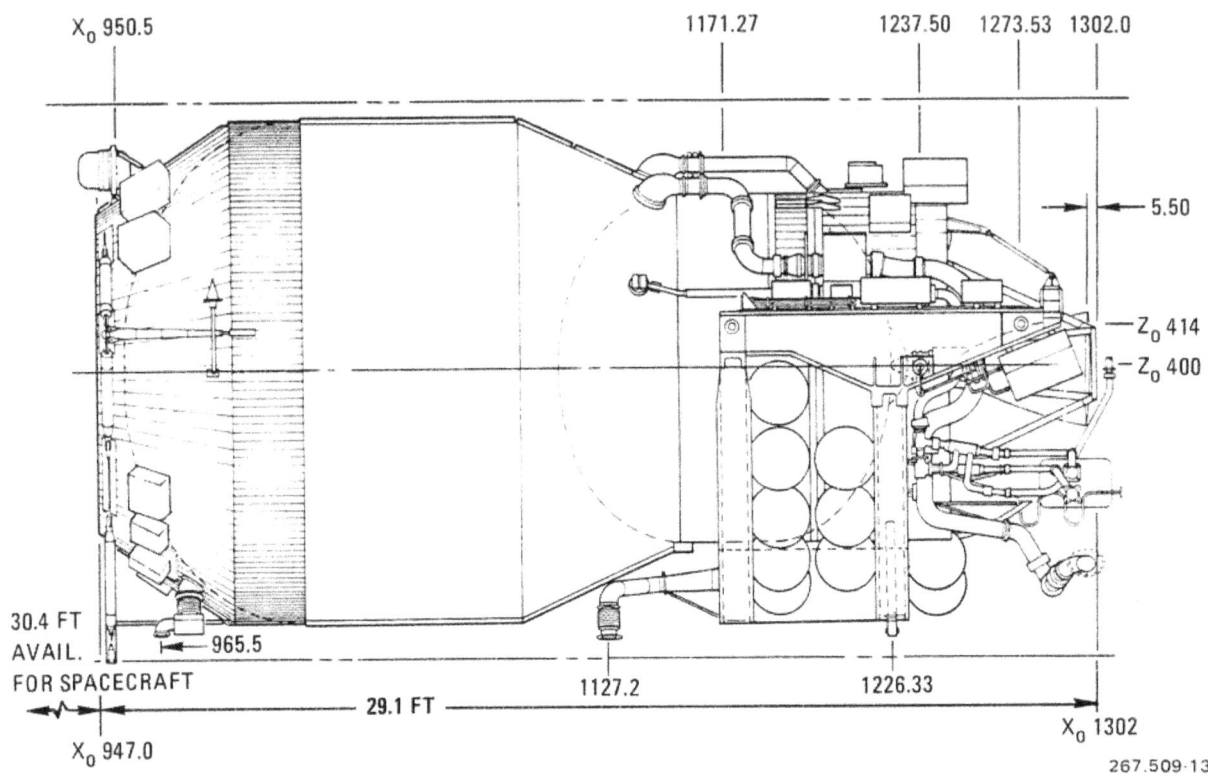

Figure 3-1. Centaur G-Prime Configuration

3.2.3 MISSION PERFORMANCE

3.2.3.1 Earth-Escape Mission. Centaur G-prime can inject 5,302 pounds of payload into an earth-escape orbit with an energy of 80 km^2/sec^2 for launches from the Eastern Launch Site using a 130-n.mi. circular, 28.5-degree inclined parking orbit supplied by the STS. The Shuttle lift requirement for GLL is 65,000 pounds and 62,500 for ISPM, as indicated by the weight summary of Table 3-2. Figure 3-2 presents the orbital energy capabilities of the Centaur G-prime for payload weights up to 6,000 lb. Mission requirements for the 1986 Galileo mission and the 1986 International Solar Polar mission are included.

3.2.4 FLEXIBILITY FOR CONTINGENCIES. The versatility of Centaur software and payload capability for existing missions provides mission flexibility allowing options for contingency operations.

Table 3-2. Centaur G-Prime Mission Weight Summaries

Item	Weight (lb)	
	Galileo	ISPM
Total loaded weight	65,000	62,449
Total support weight	9,428	9,428
STS supplied chargeables	2,900	2,900
Centaur airborne support equip	6,528	6,528
Total vehicle weight	55,572	53,021
Spacecraft weight	5,302	809
Centaur tanked weight	50,270	52,212
Centaur dry weight	6,088	5,927
Centaur residuals	600	600
Centaur expendables	43,582	45,685
Propellants (LH_2 & LO_2)*	43,531	45,633
Hydrazine	49	50
Helium	2	2

*Offloaded 2,124 lb for Galileo mission.

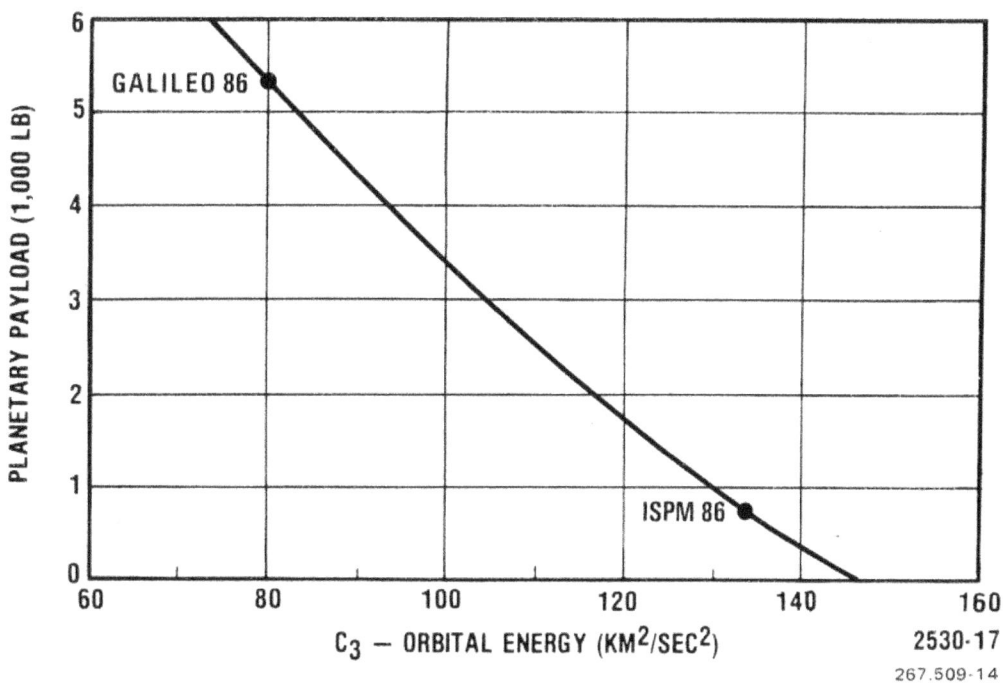

Figure 3-2. Centaur G-Prime Planetary Performance

For example, Centaur software capability increases mission flexibility by allowing Centaur deployment and/or mission initiation on successive revolutions in the parking orbit. Orbit parameters can be selected automatically from previously validated multiple-targeting sets as a function of time to account for mission initiation delays. Software for contingency options has been flight-proven; typical examples are automated inflight retargeting capability for HEAO launches (Centaur D-1A) and provision for contingency parking-orbit revolution for Voyager launches (Centaur D-1T).

3.3 CENTAUR G-PRIME CONFIGURATION

This discussion of the configuration is structured to provide a basic description of the baseline single-burn planetary mission, Shuttle/Centaur G-prime. Mission-peculiar changes dictated by unique requirements are discussed in applicable sections.

3.3.1 CENTAUR SYSTEM GENERAL CONCEPT

3.3.1.1 Centaur Vehicle. The Centaur vehicle (Figure 3-1) consists of a 10-foot-diameter LO_2 tank that transitions to an LH_2 tank with a diameter of 14 feet 2 inches. The cryogenic tanks are insulated with combinations of helium-purged foam blankets and radiation shields. The forward end of the vehicle incorporates a bolted-on cylindrical forward adapter which provides mounts for all vehicle electronics packages. The aft end of the vehicle incorporates a cylindrical aft adapter and a pyrotechnic separation ring.

The vehicle avionics system performs the functions necessary for autonomous control of the Centaur vehicle from Orbiter separation through post-separation maneuvers.

The GN&C system for the baseline Centaur G-prime is similar to the Centaur D-1A GN&C system. The system was completely redesigned to NASA Hi-Rel standards in 1972.

The telemetry system is compatible with the Tracking and Data Relay Satellite System (TDRSS) link, and permits data uplink via the Orbiter when Centaur is attached.

Electrical power to safety-related avionics control functions is inhibited until the Centaur is a safe distance from the Orbiter.

3.3.1.2 Centaur Integrated Support System (CISS). The CISS consists of a Centaur support structure (CSS), a deployment adapter, and the associated CISS electronics and fluid systems (Figure 3-3). The CSS adapts the Centaur vehicle and deployment adapter to the Orbiter through a five-point support system. The deployment adapter attaches to the aft end of Centaur at the separation ring and to the CSS through two rotation trunnions and guide keel pin.

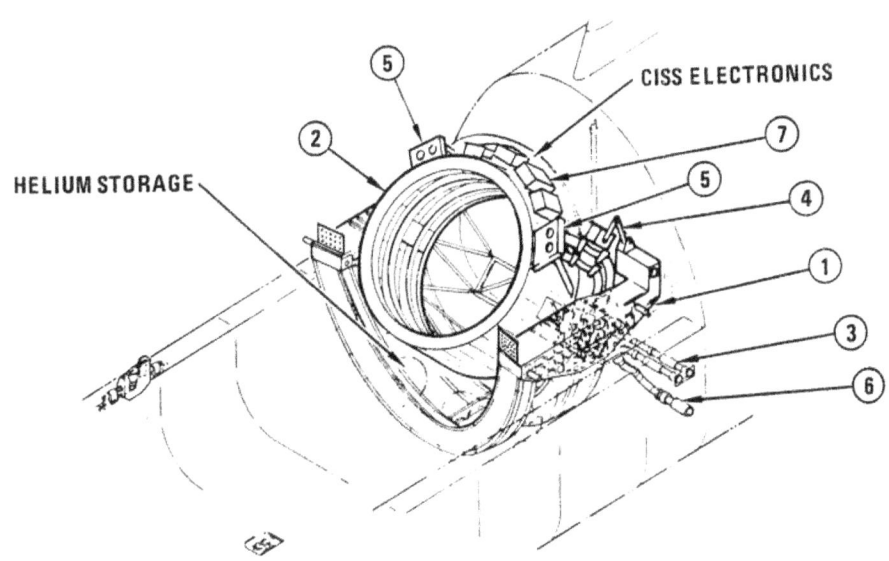

INTERFACE COMPATIBILITY

① FIVE-POINT AFT SUPPORT SYSTEM BETWEEN CISS AND ORBITER (STANDARD FITTINGS WITH ADJUSTMENT CAPABILITY ADDED TO FORWARD SILL LATCHES)

② DEPLOYMENT ADAPTER WITH SPRING THRUST TO EJECT CENTAUR

③ FLUID SERVICING FOR LH$_2$ FILL & DRAIN AND PRELAUNCH TANK VENTING

④ DEPLOYMENT ADAPTER ROTATION SYSTEM (TO 45 DEGREES)

⑤ FUEL AND OXIDIZER DISCONNECT PANELS

SAFETY CONSIDERATIONS

⑥ PROPELLANT ABORT DUMP

⑦ FIVE AUTONOMOUS CONTROL UNITS PROVIDE TWO-FAILURE TOLERANT CONTROL

267.509-17

Figure 3-3. Centaur Integrated Support System (CISS)

During deployment, the vehicle is rotated 45 degrees to its separation attitude by a rotation mechanism attached to the deployment adapter.

Fluid systems ducting and gimbals are provided to interconnect the various propellant tank service lines to their associated Orbiter overboard service ports (Figure 3-4). The gimbals permit the Centaur to be rotated to the deployment position while maintaining all safety-related systems in a connected and functional state.

Helium storage spheres, single-failure tolerant pressurization systems, and two-failure tolerant pressure regulation systems supply all helium for pressurizing Centaur tanks, actuating vent and dump system valves, and providing the necessary system purges to manage Centaur propellants safely.

CISS avionics performs all control functions for vehicle safety while the Centaur is attached to the Orbiter. Two-failure tolerant control is achieved with five strings of microprocessor-controlled avionics, associated sensors, and controllers.

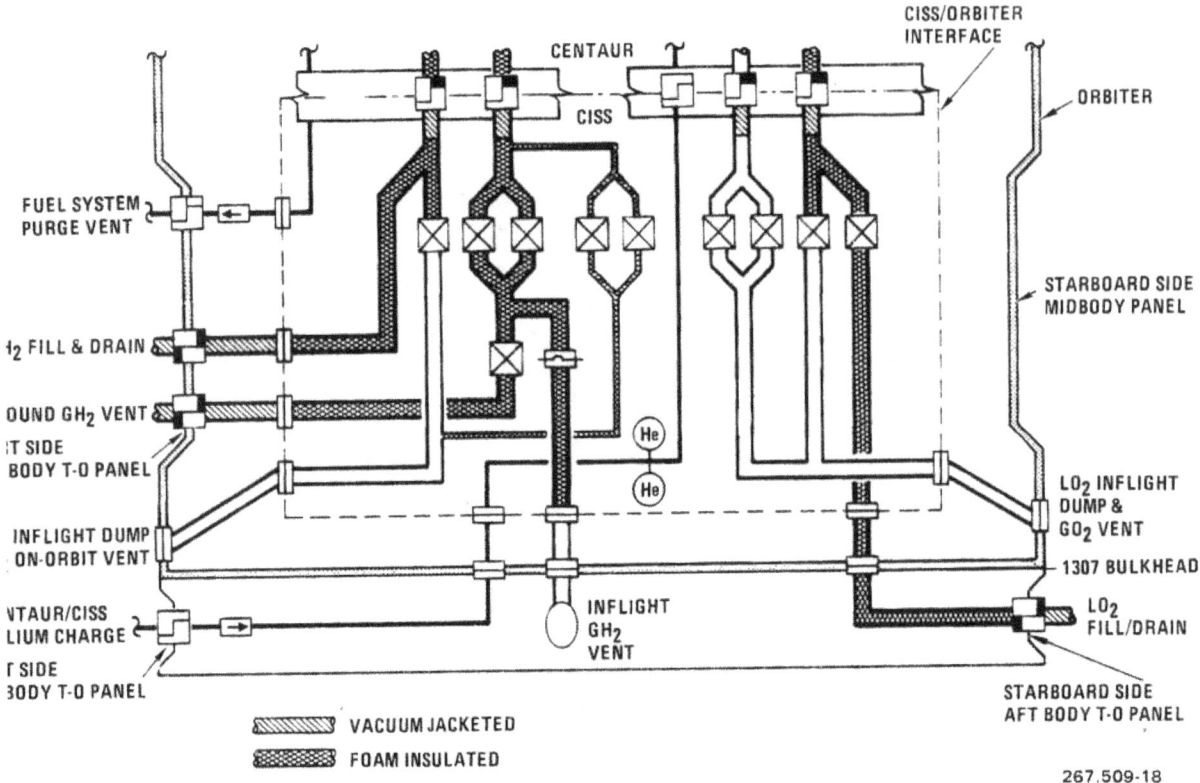

Figure 3-4. Centaur/Orbiter Fluid Interfaces

3.3.2 STRUCTURAL SYSTEMS

3.3.2.1 Centaur Vehicle. Structural components are designed to provide ultimate factors of safety equal to or greater than 1.40 while in the Orbiter bay and 1.25 after separation for all mission phases. Components are designed to the limit load factor specified in Tables 3.1.3.1 through 4.1.3.4 of JSC 07700, Vol. XIV, Attachment 1. Loads will be verified by dynamic loads analysis using Orbiter forcing functions supplied by JSC/Rockwell.

Structural components are designed to react emergency landing loads using ultimate design factors for nonreturnable payloads given in Table 4.1.3.5-1 of JSC 07700, Vol. XIV, Attachment 1.

a. Centaur Tank Configuration. The basic propellant tank arrangement is a LO_2 tank and a LH_2 tank, as illustrated in Figure 3-5. The weight-effective, pressure-stabilized tank configuration has been proven in 463 Atlas flights, 61 Centaur D-1A flights, and 7 Centaur D-1T flights. This structurally efficient Centaur G-prime tank contains the main engine propellants, establishes vehicle primary structural integrity, and supports vehicle systems and components. Extensive testing of the Centaur D-1A tank has demonstrated the tank has a much greater strength capability than the design values.

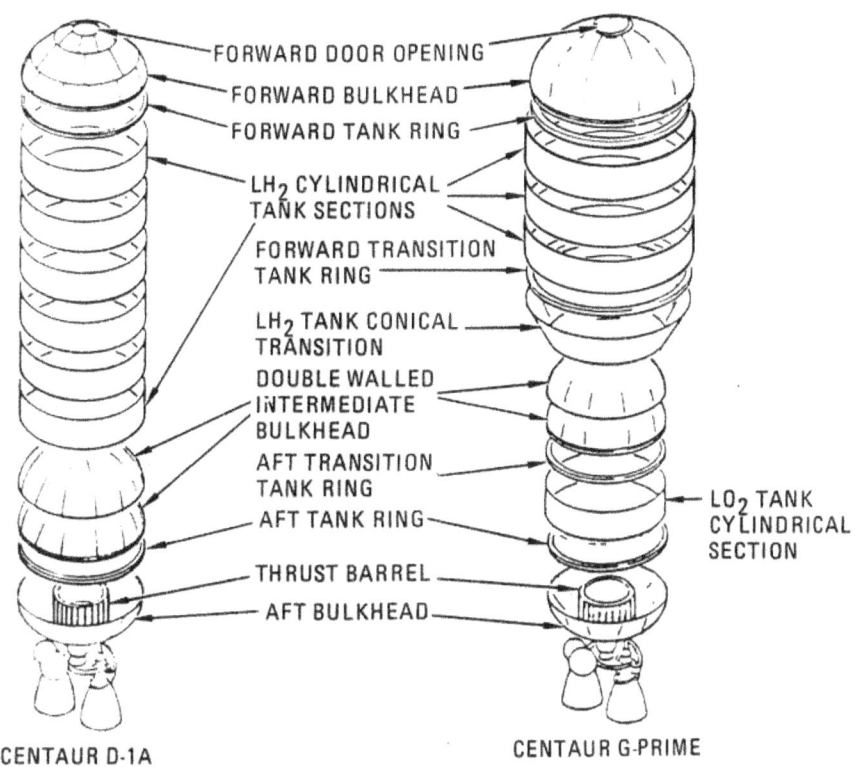

FORWARD DOOR OPENING

FORWARD BULKHEAD

FORWARD TANK RING

LH$_2$ CYLINDRICAL TANK SECTIONS

FORWARD TRANSITION TANK RING

LH$_2$ TANK CONICAL TRANSITION

DOUBLE WALLED INTERMEDIATE BULKHEAD

AFT TRANSITION TANK RING

AFT TANK RING

THRUST BARREL

AFT BULKHEAD

LO$_2$ TANK CYLINDRICAL SECTION

CENTAUR D-1A

CENTAUR G-PRIME

267.509-19

Figure 3-5.   Centaur D-1A and Centaur G-Prime Tanks

The basic tank material is 301 CRES. The entire welded tank assembly,
including all rings and brackets, is made of 300 series CRES, which
minimizes galvanic corrosion. Stress corrosion cracking problems are
avoided by the selection of the tank materials and by storage and
maintenance activities that provide a periodic protective coating.

Tank raw stock is ultrasonically inspected. Tensile, elongation, and weld
joint fatigue tests are run on all tank raw stock and this procedure will
be continued. Major structural member integrity is verified by test
coupons from parent material. Tank weld samples are tested before,
during, and after the machine welding operation. This existing procedure
will be continued. All tank weld joints are 100% radiographically
inspected. Leak testing is performed on all tank weld joints. All
existing and proven Centaur reliability and quality assurance requirements
will continue to be imposed on the tank design and manufacturing.

The LO$_2$ tank is formed by two ellipsoidal bulkheads of 120-inch major
diameter and 87.0-inch minor diameter the same as the Centaur D-1A tank
with a 31-inch cylindrical section inserted between the bulkheads.

The LH$_2$ tank will consist of a 170-inch-diameter cylindrical section
closed by an ellipsoidal forward bulkhead and a 24-degree conical aft
transition bulkhead that attaches to the LO$_2$ tank at its forward
bulkhead/cylindrical section joint (Figure 3-6).

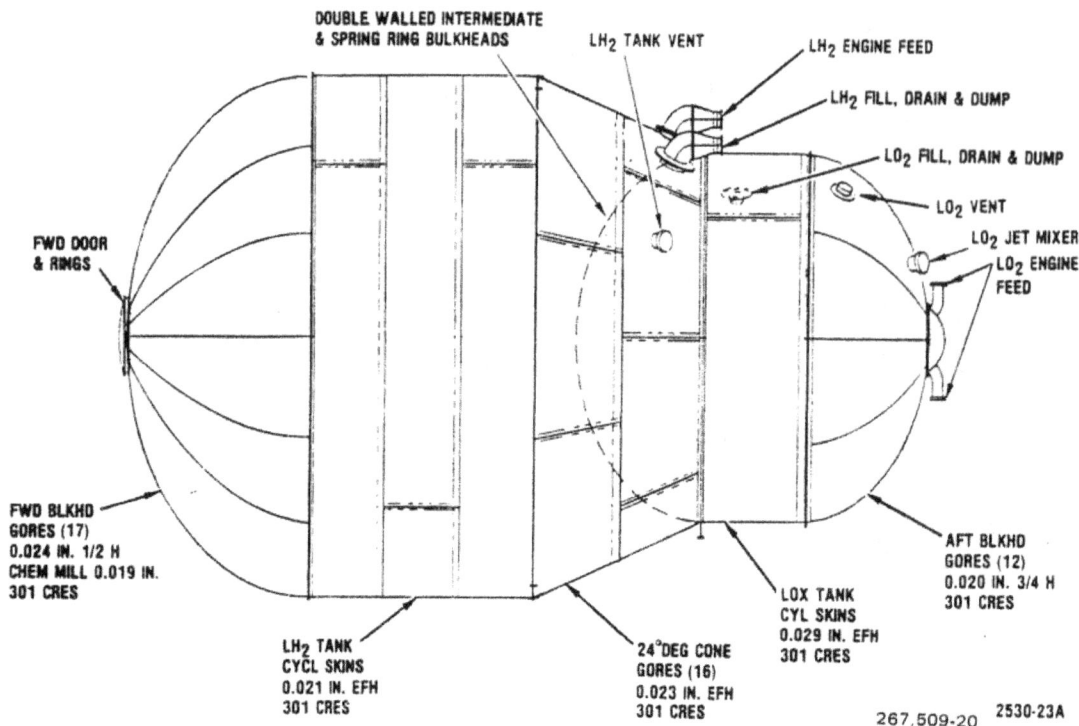

Figure 3-6. Centaur G-Prime Propellant Tank Configuration

The 170-inch diameter LH_2 tank cylinder provides adequate thermal and dynamic clearance between the Centaur G-prime and the STS Orbiter payload envelope. Structural rings, insulation, and zero-g vent duct space requirements are accounted for as shown in Figure 3-7.

The tank volumes are designed to provide total propellants of 46,000 lb at a burn mixture ratio of 5.0 pounds (LO_2) to 1.0 pound (LH_2). Tank skin gages are determined by internal pressures resulting from Centaur engine inlet pressure requirements (see Figures 3-8 through 3-11).

The LH_2 tank forward bulkhead consists of 17 gore sections buttwelded to form an ellipsoid, as shown in Figure 3-8. Except for the buttweld land areas, the gores are chem-milled to reduce weight. At the forward end of the bulkhead, an access door is bolted to a door ring, which is welded to the bulkhead. Welded to the inside of the door ring is support structure for the tank zero-g vent system, level sensors, and instrumentation. Attachment brackets for various items, such as the zero-g vent pressure lines, wiring, and insulation, are welded to the external surface. Welded to the interior of the bulkhead is attachment bracketry for the pressurizing gas diffuser/ dissipator and the zero-g vent duct as shown in Figure 3-11.

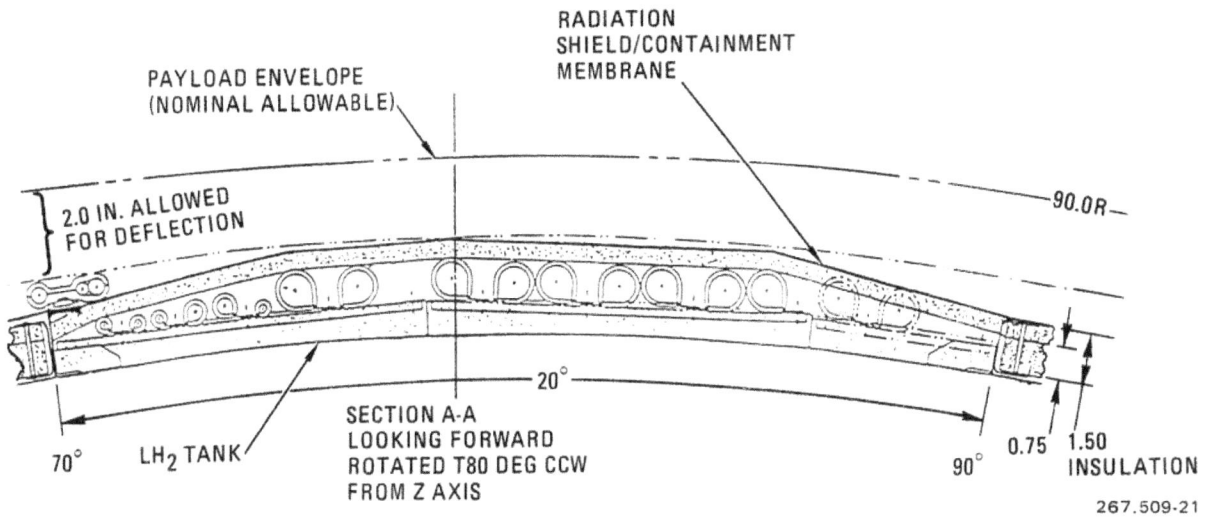

SECTION A-A
LOOKING FORWARD
ROTATED T80 DEG CCW
FROM Z AXIS

267.509-21

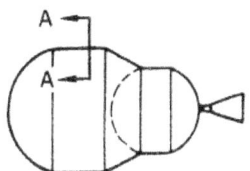

Figure 3-7. 170-Inch-Diameter Centaur G-Prime LH$_2$ Tank

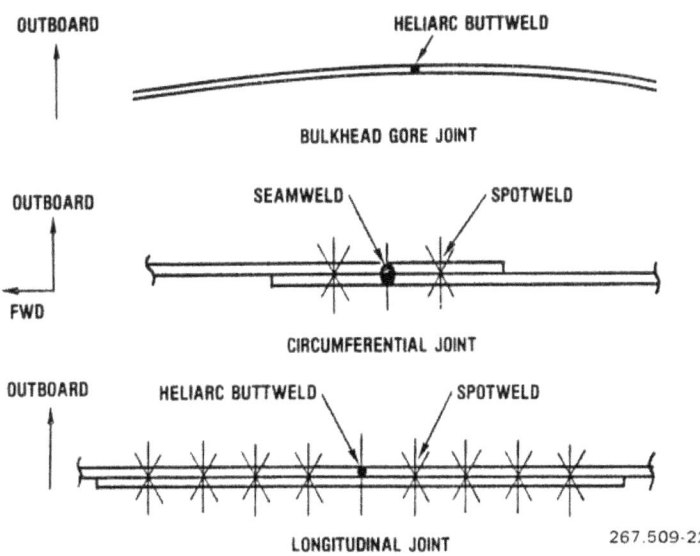

267.509-22

Figure 3-8. Typical Centaur G-Prime and Centaur D-1A Tank Joints

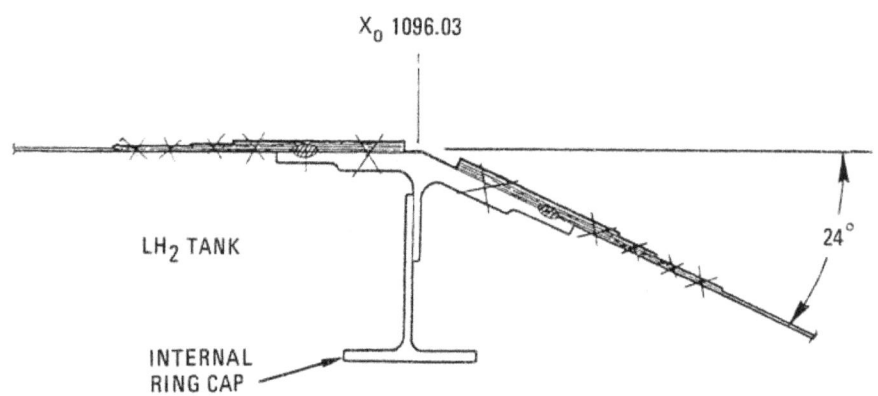

X_0 1096.03

LH$_2$ TANK

24°

INTERNAL
RING CAP

LH$_2$ CYLINDER-TO-CONE TRANSITION JOINT

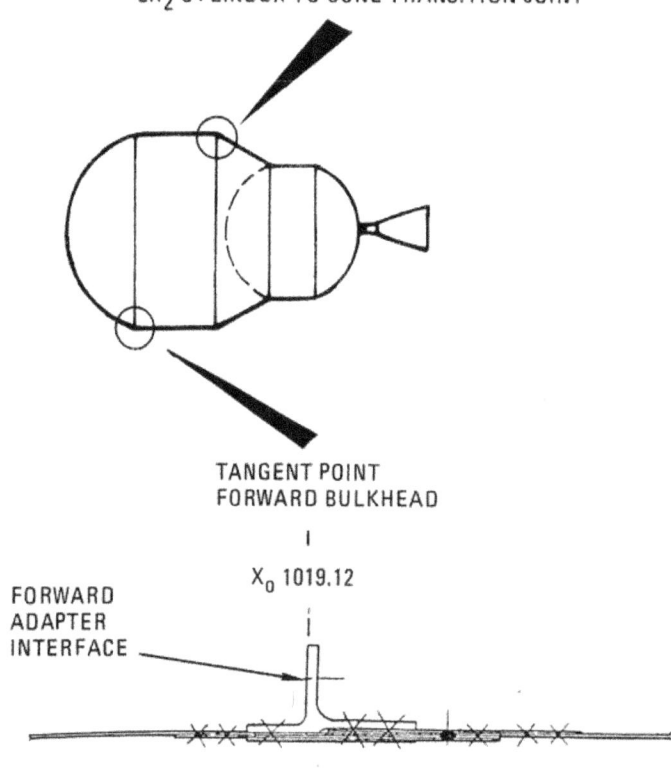

TANGENT POINT
FORWARD BULKHEAD

X_0 1019.12

FORWARD
ADAPTER
INTERFACE

FORWARD BULKHEAD TO LH$_2$ CYLINDER TRANSITION JOINT

267.509-23

Figure 3-9. Hydrogen Tank Transition Joints

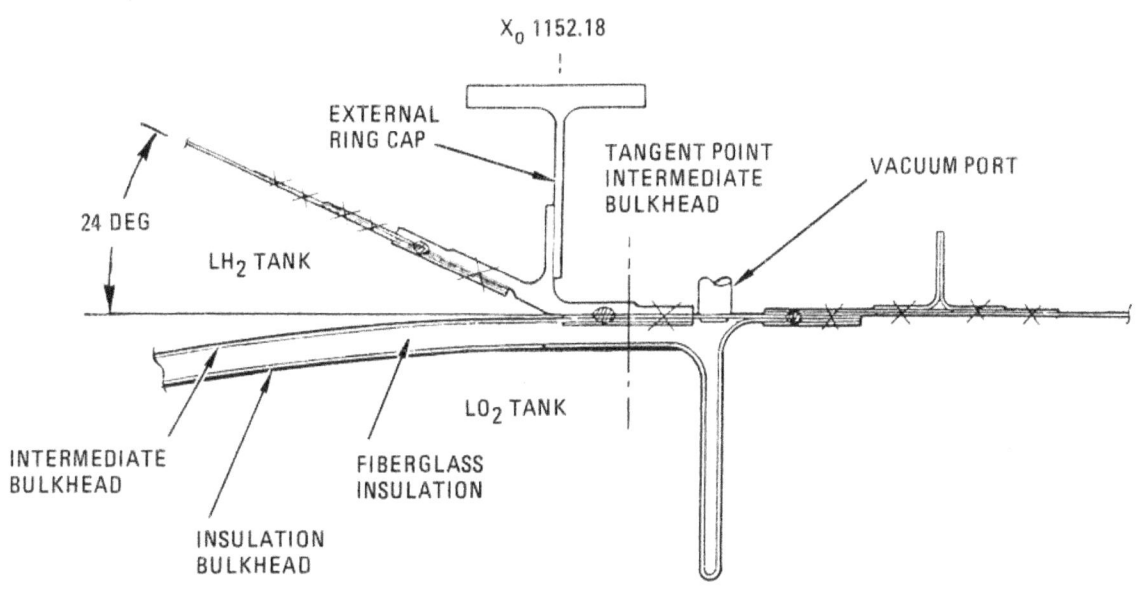

LH$_2$ CONE TO LO$_2$ CYLINDER TRANSITION JOINT

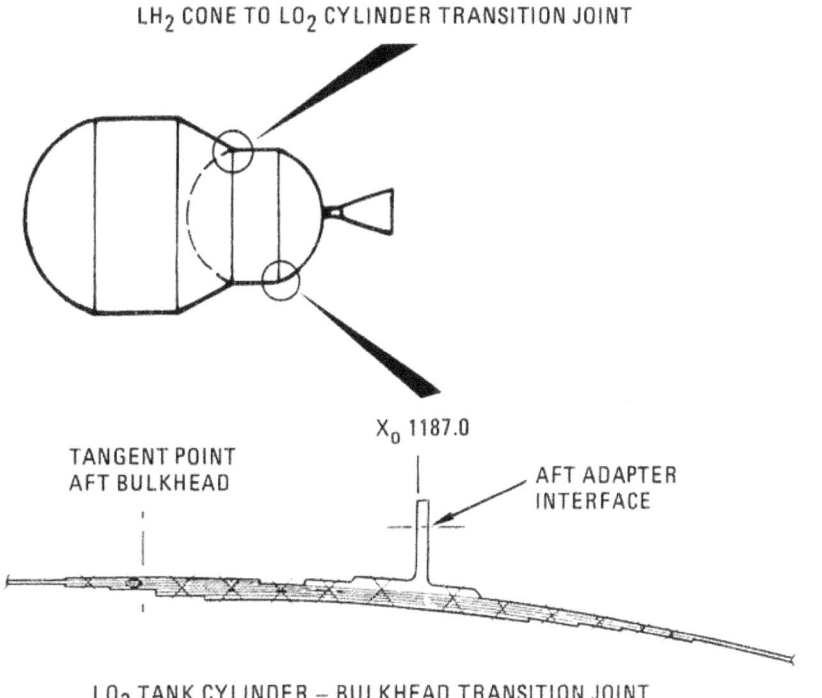

LO$_2$ TANK CYLINDER — BULKHEAD TRANSITION JOINT

267.509-24

Figure 3-10.   Oxygen Tank Transition Joints

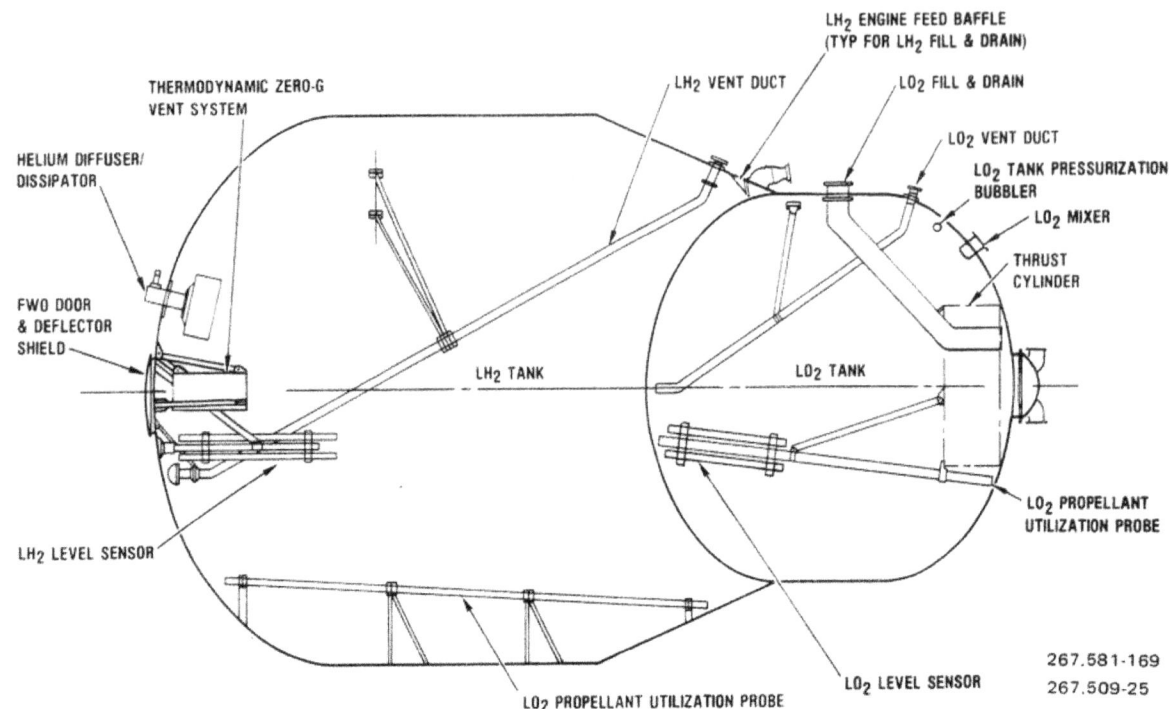

Figure 3-11. Centaur G-Prime Tank Internal Installations

The LH$_2$ tank cylindrical section consists of three sections about 25 inches long and 170 inches in diameter. The three skins are spotwelded, seamwelded, and stove-piped together using existing production techniques as shown in Figure 3-8. The forward skin mates with the aft end of the forward bulkhead and will contain the forward adapter support ring. This ring is shown in Figure 3-9. The aft skin mates to the LH$_2$ tank aft cone transition as shown in Figure 3-9. Both joints use existing Centaur D-1A type spot/seamweld joints. Support structure for items such as vent, pressure lines, purge lines, wiring, and insulation is welded to the external surface.

The LH$_2$ tank aft-cone section will consist of sixteen 24-degree conic skin gores buttwelded together and attached to transition rings at each end. The aft transition ring is also attached to the LO$_2$ tank at the forward end of its cylindrical section. The aft cone LH$_2$ section will contain LH$_2$ propellant feed, fill and dump, and vent standpipe outlets as shown in Figure 3-6. Support structure for various items, such as the vent and pressure lines, wiring, and insulation is welded to the external surface.

The LO_2 tank forward structural bulkhead is identical to the existing Centaur D-1A. It consists of gore sections buttwelded together to form an ellipsoid and is attached to the forward end of the LO_2 tank cylindrical section. The gore sections are chemically milled to reduce weight except in areas of the buttwelds (refer to Figures 3-5 and 3-10). The structural bulkhead will be spot and seamwelded to the aft transition ring providing the same sealing and structural integrity as in the existing Centaur D-1A.

The LO_2 tank forward spring ring/insulation bulkhead is identical to the existing Centaur D-1A assembly except the aft flange of the spring ring will attach to the LO_2 tank cylindrical section. The insulation bulkhead consists of gore section buttwelded together to form an ellipsoid and is buttwelded to the formed spring ring. The cavity between the forward structural bulkhead and insulation bulkhead contains insulation identical to the existing Centaur D-1A (see Figures 3-5 and 3-10).

The LO_2 tank cylindrical section consists of one section about 30 inches long and 120 inches in diameter. The forward end is welded to the LH_2 aft cone/LO_2 forward bulkhead/spring ring transition joint shown in Figure 3-10. The aft end is welded to the LO_2 tank aft bulkhead. This joint also contains the aft tank ring on the aft bulkhead near the cylindrical section of the LO_2 tank shown in Figure 3-10. The ring is used to attach the aft adapter. The LO_2 tank cylindrical section contains the fill/dump siphon duct and outlet elbow terminating in a fill and drain valve mounted to the skin. Support structure for various items, such as vent and pressure lines, wiring, and insulation, is welded to the external surface (see Figure 3-11).

The LO_2 tank aft bulkhead structure is identical to that of existing Centaur D-1A. It consists of twelve gore sections buttwelded together to form an ellipsoid as shown in Figure 3-6. Support structure is welded to the bulkhead to support propellant lines, helium, and hydrazine bottles, wiring, the engines, electrical boxes, radiation shields, jet mixer, etc. Minor differences in these items are required from the existing vehicle. Internally, the bulkhead contains the LO_2 tank pressurizing bubbler ring, fill and drain duct, LO_2 standpipe, P/U probe, jet pulse mixer (JPM), and engine thrust barrel. The LO_2 propellant feed sump is mounted on the aft end of the bulkhead (Figure 3-11).

The thrust barrel structure inside the LO_2 tank is identical to the existing Centaur D-1A structure, with the exception of some support structure details.

The thrust barrel is a 50-inch-diameter, 15.5-inch-high cylinder of skin-stringer construction. It reacts engine thrust loads and distributes them into the LO_2 tank aft bulkhead. The forward ring and thrust longerons are 2124 aluminum alloy; the skin and stringers are 2024 aluminum alloy. The cylinder attaches mechanically to the internal aft ring, which in turn is welded to the LO_2 tank aft bulkhead.

b. Centaur Adapters. Two new vehicle adapters have been designed for Centaur G-prime, as shown in Figure 3-12. They are similar to existing Centaur Adapters in form and function. Extensive use is made of finite-element analysis (which has shown a very close correlation with test results) for sizing and designing these adapters.

(1) Forward Adapter. The major elements of the forward adapter are a 47.12-inch-long 56.7-degree conical section and a 25-inch-long, 170-inch-diameter cylindrical section. The conical structure is conventional aluminum skin frame stringer design. The cylindrical structure section is corrugated graphite/epoxy. The conical and cylindrical structures interface through a common, integrally machined, 170-inch-diameter transition ring.

The attached forward support system uses titanium sill and keel trunnion beams with boron aluminum struts. The struts have titanium end fittings with adjustment capabilities. The forward support structure attaches to the conical structure through integrally machined aluminum fittings. Centaur equipment, chiefly avionics, is mounted on the conical surface. This conical section includes:

- Two deployable antennas that will be spring-deployed after a safe separation distance from the Orbiter is achieved.

- A stable mounting platform for the inertial reference unit.

- A gas-conditioning duct for ground cooling of equipment (Figure 3-15).

- A purge vent door that will open for Orbiter ascent and close for abort return, as required.

This beam and strut support system interfaces with the Orbiter midfuselage payload support system at three points and provides forward support for Centaur in the Orbiter.

The payload mission-peculiar adapters will interface with the forward end of the forward adapter (see Figure 3-13).

(2) Aft Adapter. The aft adapter (Figure 3-14) is a 10-foot-diameter, 11.2-inch-long, graphite/epoxy composite structure with attachment rings at each end. This adapter distributes CISS support loads into the Centaur tank and provides an interface for attaching the separation system. The forward ring bolts to the LO_2 tank aft ring and the aft ring attaches to the separation ring. The aft adapter is similar in design to the proven Centaur D-1A interstage adapter. Common design features include attachment ring configuration and bolt pattern and cutout locations. The skin is of variable thickness. The adapter cutouts are designed for routing the LH_2 propellant feed duct and electrical wiring from the LO_2 tank cylindrical section to the LO_2 tank aft bulkhead area, and for routing small tubing. Light, efficient support structure is mounted on the aft adapter for the vehicle separation spring fittings, fluid and electrical disconnect panels, radiation shields, and wiring.

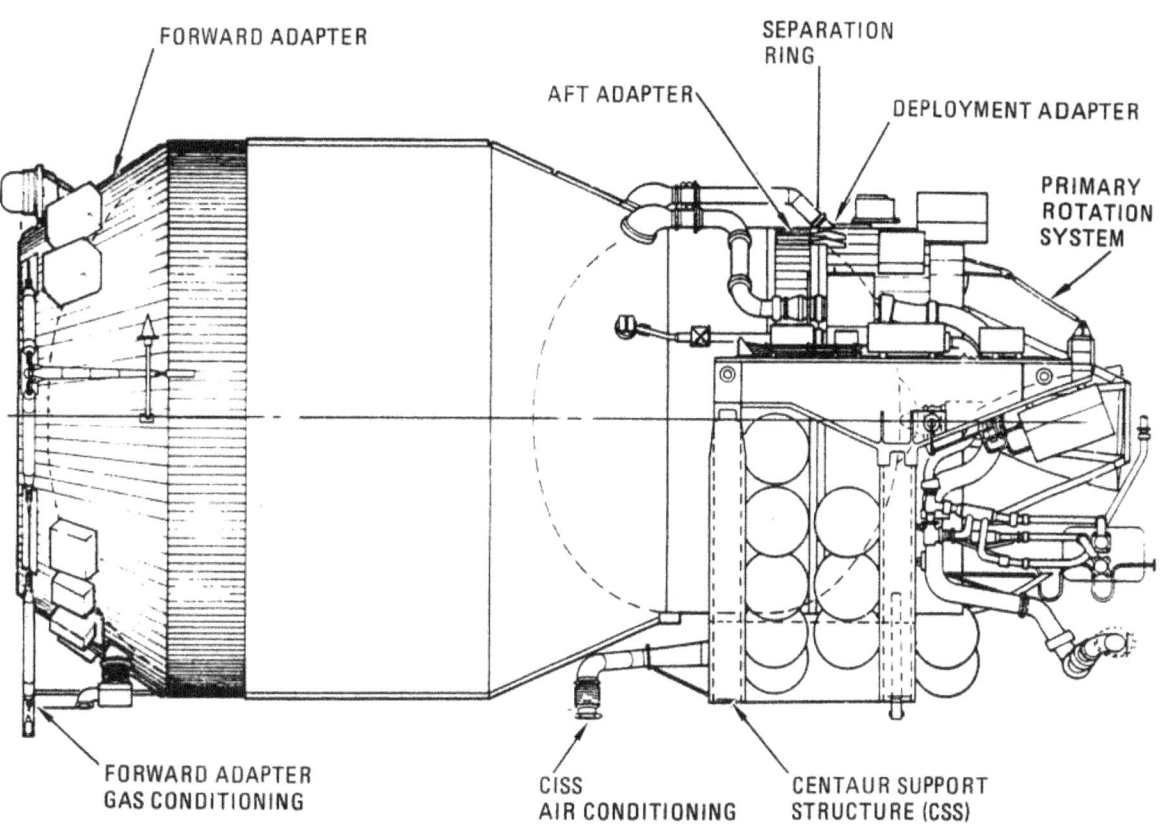

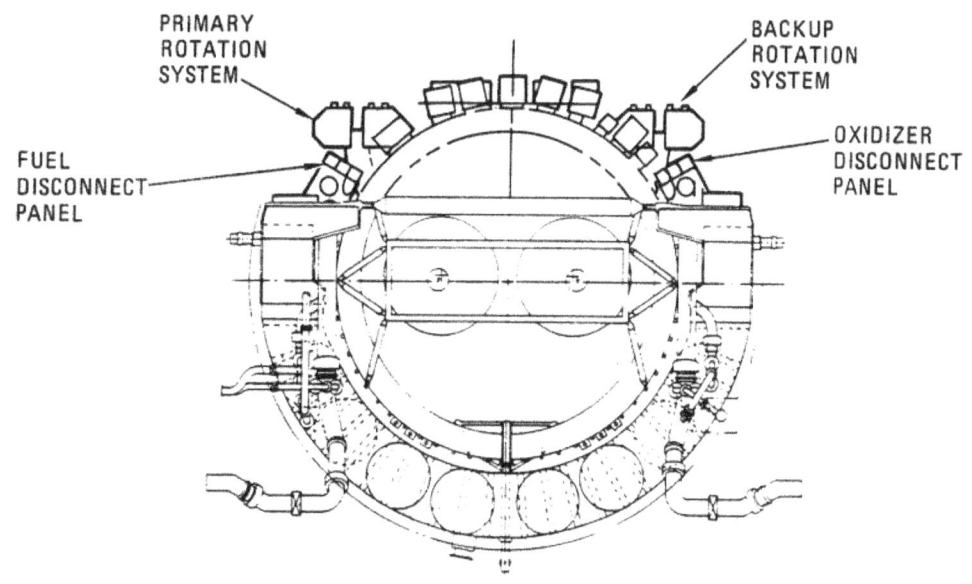

267.509-27

Figure 3-12.   Centaur G-Prime Adapters

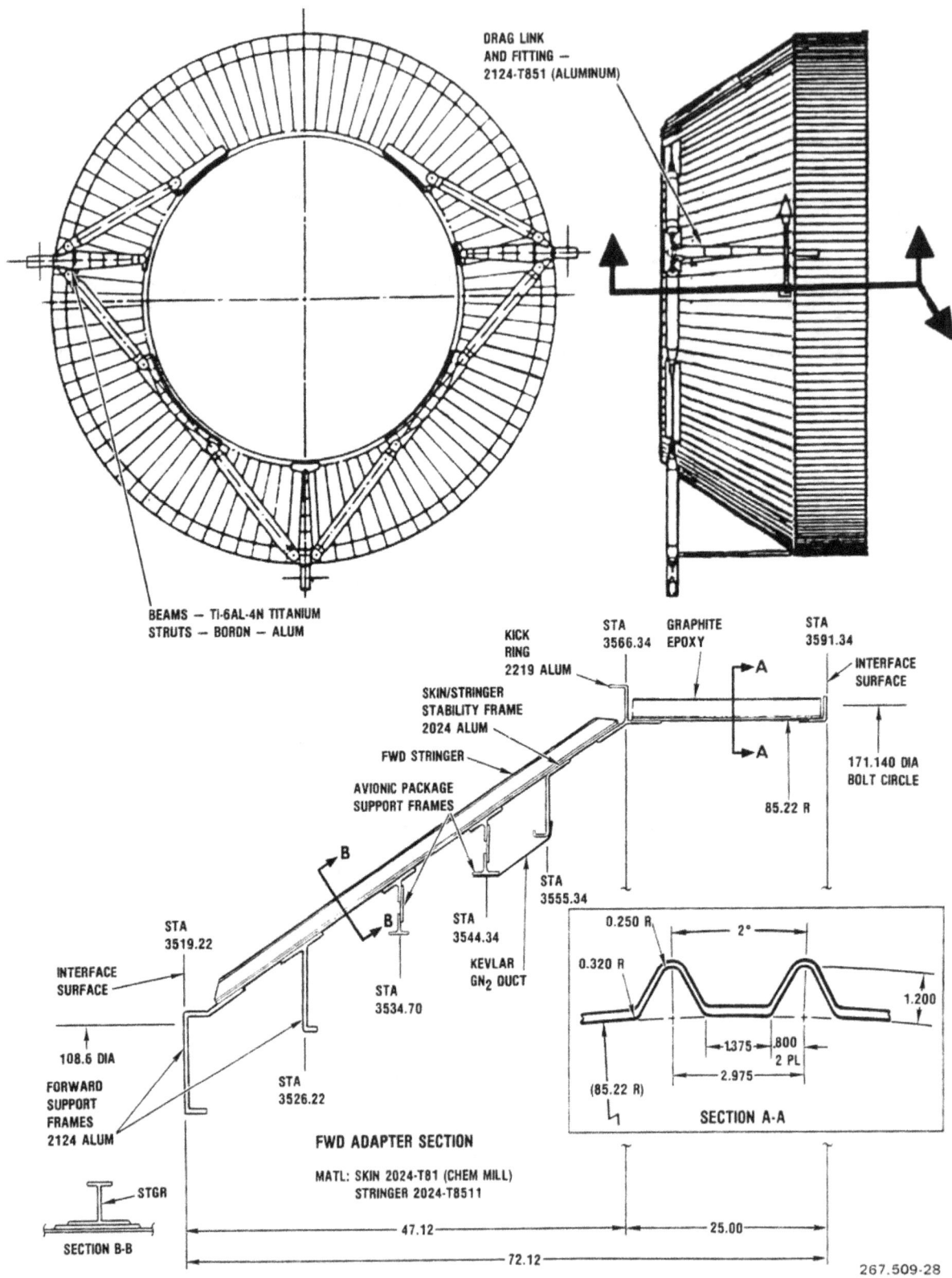

Figure 3-13. Forward Adapter

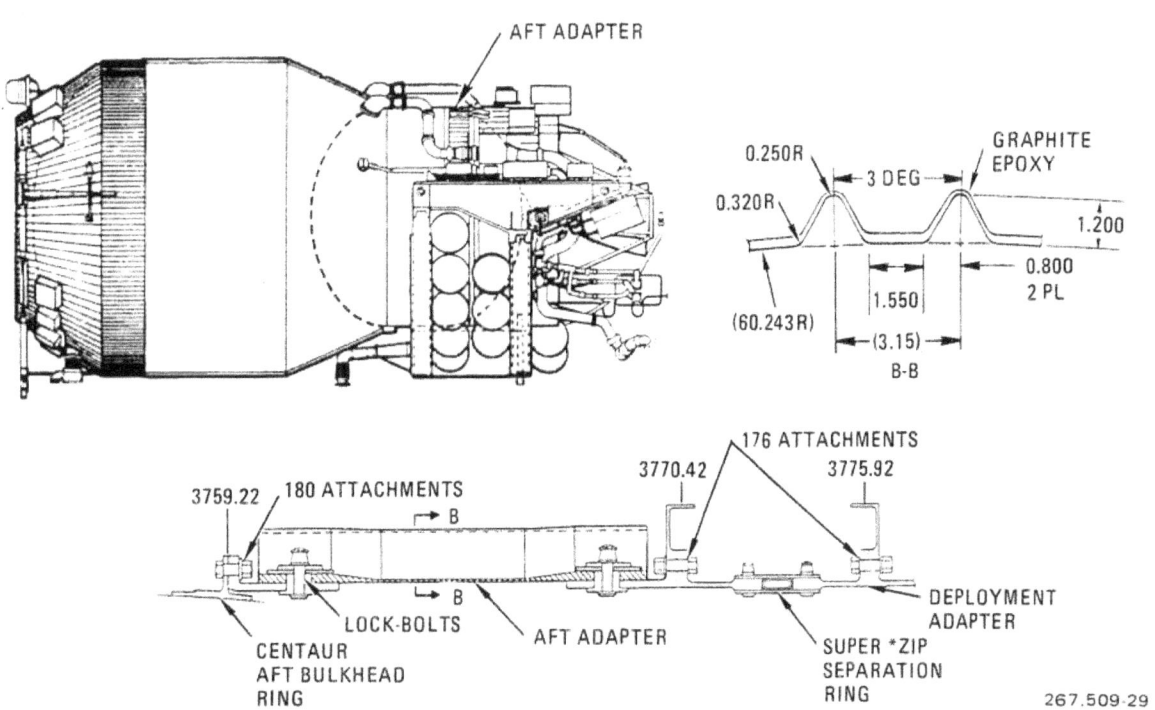

Figure 3-14. Aft Adapter

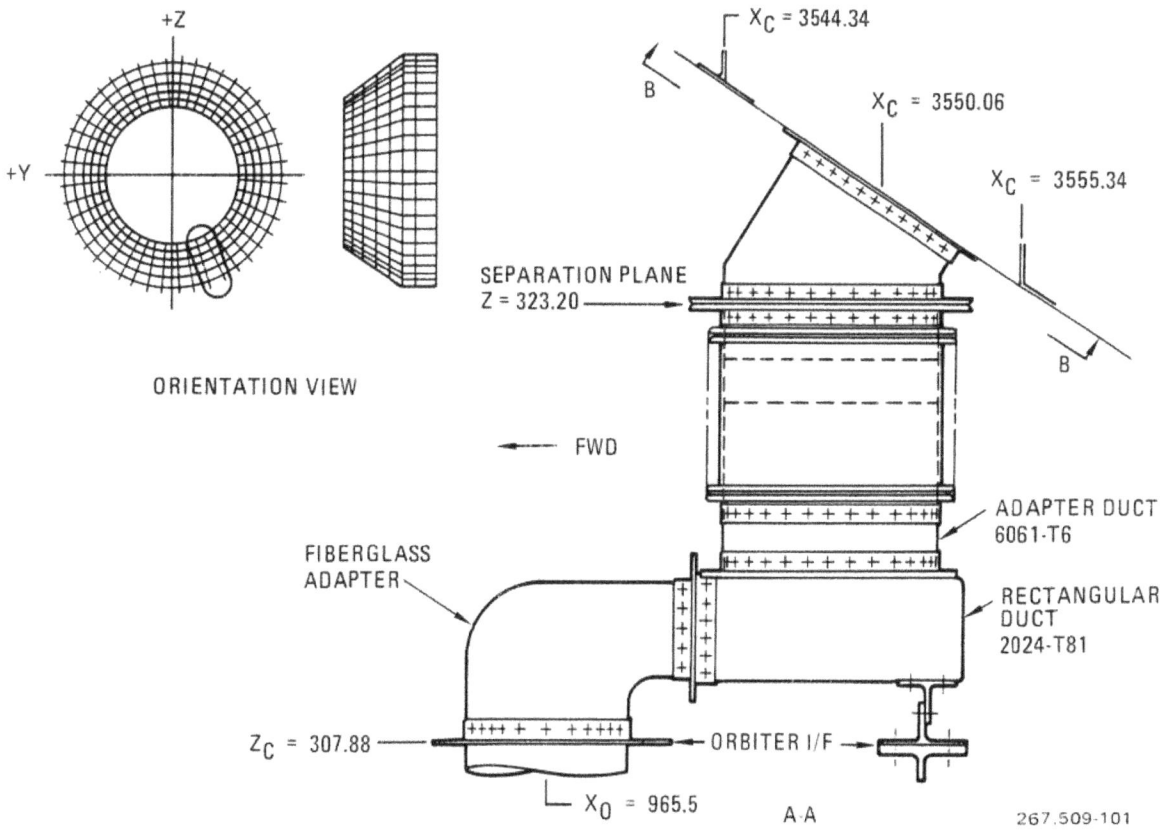

Figure 3-15. Forward Air Conditioning Duct Interface

c.   Separation System.   The Lockheed Super*Zip pyrotechnic separation system
     is used to separate Centaur from the CISS (and Orbiter).   Lockheed
     provides a separation ring containing the Super*Zip system.   It is a
     10-foot-diameter, 5.50-inch-long, aluminum-alloy cylinder section with
     attachment rings at each end.   The separation ring simply bolts to the aft
     adapter and the CISS deployment adapter.

     Super*Zip is a dual pyrotechnic system.   When it fires, a spring system
     thrusts the Centaur from the CISS deployment adapter.   Should the
     Super*Zip pyros not fire, the Centaur and deployment adapter can safely be
     lowered back into the payload bay, thus providing a two-failure tolerant
     system.   A Super*Zip system was used for shroud separation on Titan/
     Centaur.

d.   Insulation System.   The Centaur G-prime insulation system is functionally
     similar to the Centaur D-1A forward bulkhead insulation system.   All
     insulation materials meet STS contamination and safety requirements.

     (1) LH$_2$ Tank Insulation System.   This system consists of two major
          portions (Figure 3-16):   the forward bulkhead insulation, and the tank
          sidewall insulation.   The forward bulkhead two-layer foam insulation
          blankets are installed on the hydrogen tank forward bulkhead covered
          with three radiation shields and enclosed by the forward adapter and
          the purge containment diaphragm.   The tank sidewall two-layer foam
          insulation blankets are attached at the forward tank ring and extend
          aft along the full length of the hydrogen tank sidewall cylindrical
          and conical section and are attached along the cableway and to the
          plenum support angle.

          Sidewall insulation blankets are enclosed by three radiation shields
          that are sealed to the forward adapter transition ring flange and the
          purge plenum at the aft end, which provides containment of the blanket
          helium purge gas.   A vent door is provided in the forward adapter
          compartment during ascent.   The insulation blanket compartments are
          purged with helium before tanking to purge the GN$_2$ from the blanket.
          During tanking, this helium purge keeps out the payload bay GN$_2$
          purge and maintains a positive ΔP across the radiation shield to pro-
          vide insulation.   In the event of an abort, the purge is reactivated
          to preclude moisture and air from entering during the abort descent
          and after touchdown.   The purge gas enters the forward blanket com-
          partment through a purge tube in the forward adapter and flows over
          and through the forward bulkhead blanket, through the forward adapter
          cylindrical skin and aft along the sidewall blanket into the purge
          plenum.

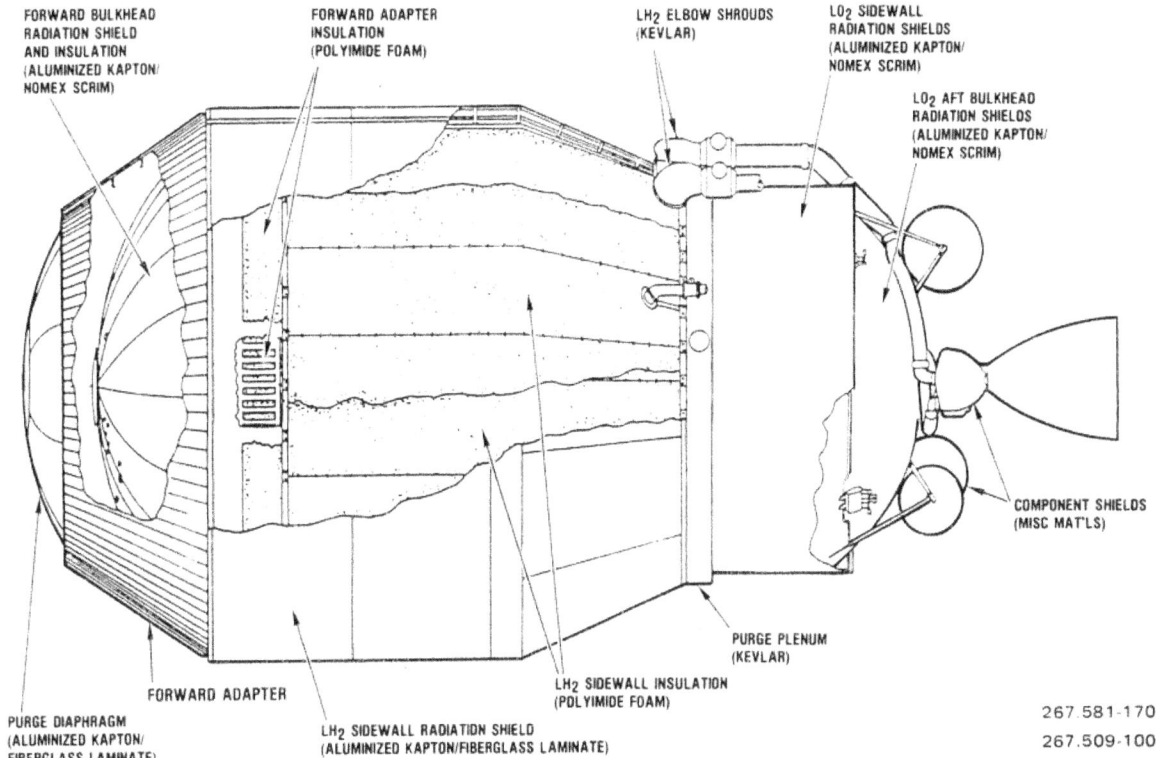

FORWARD BULKHEAD
RADIATION SHIELD
AND INSULATION
(ALUMINIZED KAPTON/
NOMEX SCRIM)

FORWARD ADAPTER
INSULATION
(POLYIMIDE FOAM)

LH₂ ELBOW SHROUDS
(KEVLAR)

LO₂ SIDEWALL
RADIATION SHIELDS
(ALUMINIZED KAPTON/
NOMEX SCRIM)

LO₂ AFT BULKHEAD
RADIATION SHIELDS
(ALUMINIZED KAPTON/
NOMEX SCRIM)

COMPONENT SHIELDS
(MISC MAT'LS)

PURGE PLENUM
(KEVLAR)

FORWARD ADAPTER

PURGE DIAPHRAGM
(ALUMINIZED KAPTON/
FIBERGLASS LAMINATE)

LH₂ SIDEWALL RADIATION SHIELD
(ALUMINIZED KAPTON/FIBERGLASS LAMINATE)

LH₂ SIDEWALL INSULATION
(POLYIMIDE FOAM)

267.581-170
267.509-100

Figure 3-16.   Insulation System for Centaur G-Prime

The LH₂ tank forward bulkhead is protected/shielded first by three
radiation shields and secondly by a two-layer foam blanket. The three
radiation shields have the same properties as the middle shield of the
LH₂ sidewall insulation system, except the inner shield edge vented
(see Figure 3-17). This shield is held in place by the same pins
holding the foam gores in position.

The insulation blankets shown in Figure 3-17 are installed as two
separate blankets, each 3/4-inch thick and installed one over the
other with the butt joints offset half the width of the gore,
providing a total assembly thickness of 1-1/2 inches.

Each blanket is fabricated in 16 separate gores; they are fastened to
each other with pin fasteners. A low-level helium purge is employed
to maintain a prelaunch helium environment forward of the bulkhead as
shown in Figure 3-18, and the effective thermal conductivity of the
insulation approaches that of quiescent helium. However, early
inflight evacuation of the liberally vented insulation yields a system
of three radiation shields for on-orbit thermal control of the forward
bulkhead.

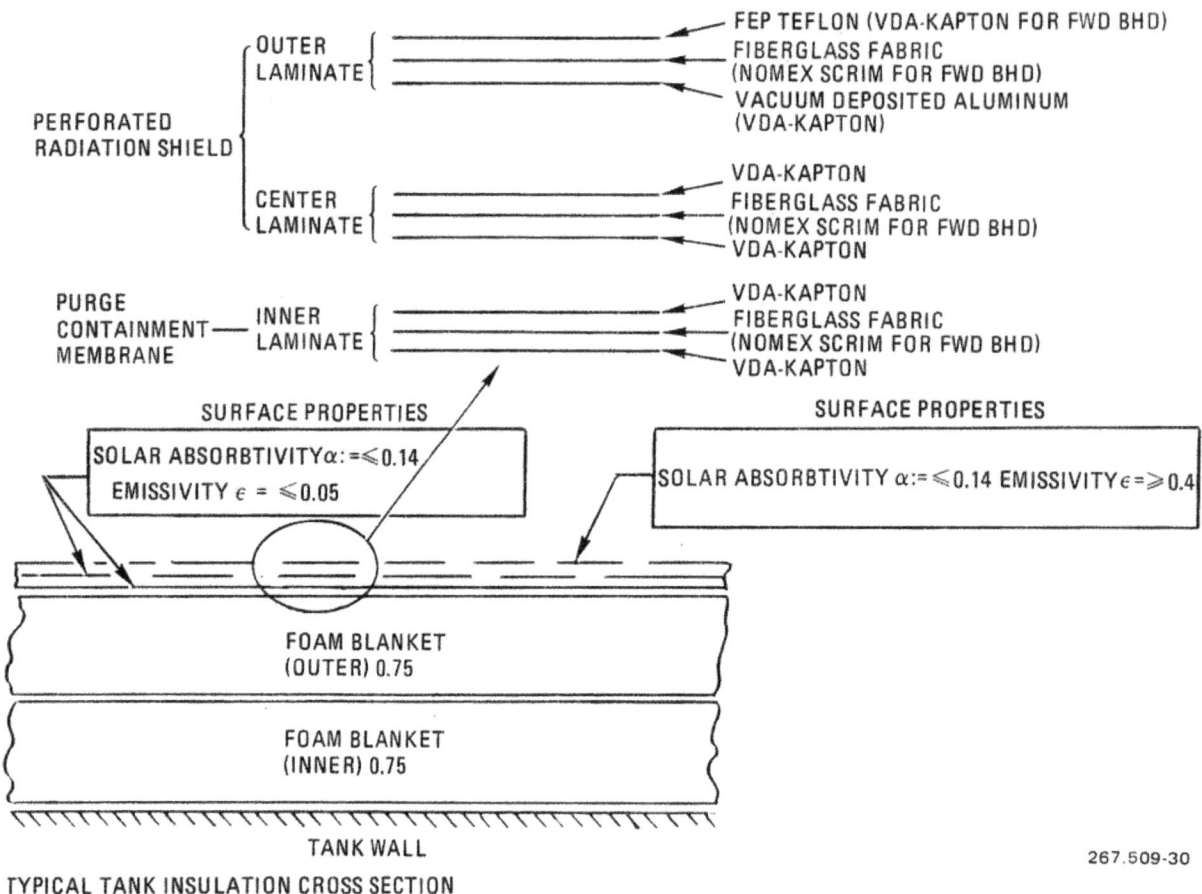

TYPICAL TANK INSULATION CROSS SECTION

267.509-30

Figure 3-17. Centaur G-Prime LH$_2$ Tank Sidewall Insulation Blankets

The foam insulation design on the Centaur forward bulkhead is also
used on the Centaur G-prime LH$_2$ tank sidewall for prelaunch thermal
control. As shown in Figure 3-17, a Kapton/glasscloth/Kapton laminate
containment membrane maintains internal helium in the blanket during
prelaunch and abort purging. Predicted prelaunch heat flux through
the LH$_2$ tank sidewall is approximately 140 Btu/hr-ft^2. Two radiation shields are positioned outboard of the helium-purged blanket.
These laminated shields are liberally ventilated to achieve rapid
inflight radiation shielding of the LH$_2$ tank sidewall. This system
of radiation shields has been thoroughly tested and corroborated on
Centaur D-1T missions.

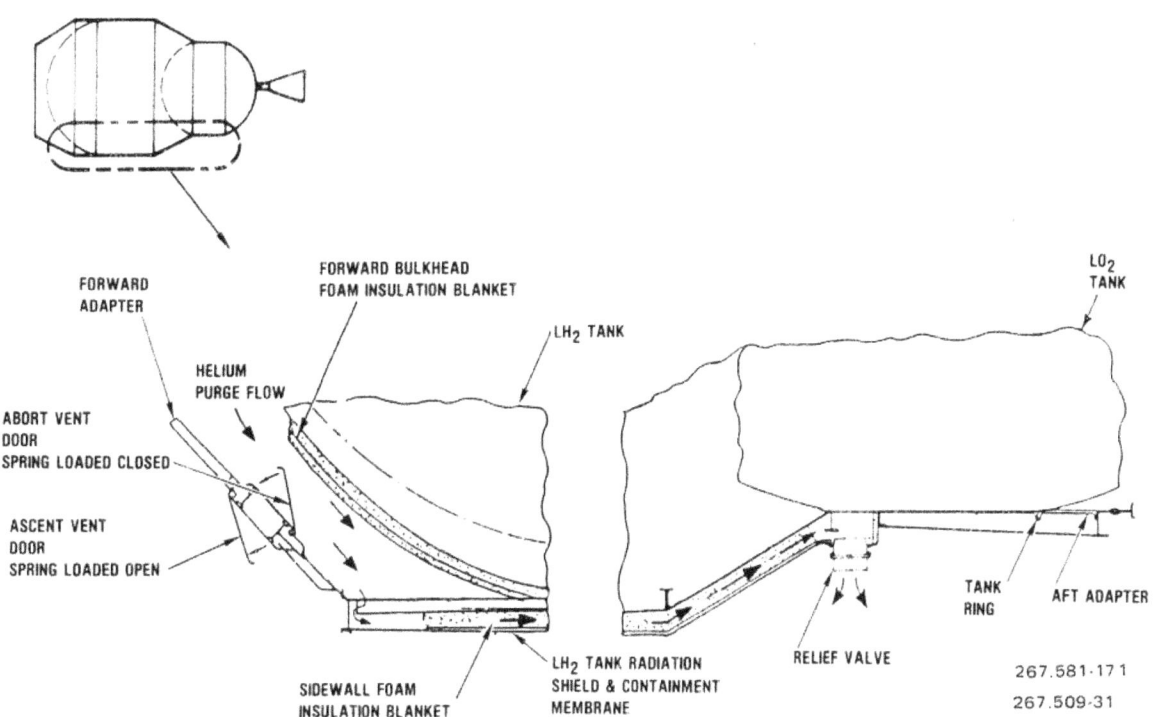

Figure 3-18. Insulation Purge and Vent Systems

(2) <u>Purge and Vent</u>. The insulation system is purged during prelaunch
conditions with helium gas introduced at ambient temperature and at up
to 160 pounds per hour at the forward end of the forward adapter. A
dual-position vent door on the forward adapter opens to vent the
compartment during ascent (Figure 3-18). Before riseoff, forward
adapter purge gas flows into the sidewall insulation, the forward
adapter, and aft to an annular purge plenum. The door is closed to
permit blanket repressurization during an abort reentry sequence when
the Centaur tank will contain postdump residual propellants. The
restart of purge flow during reentry and the foam blanket rigidity
prevents liquid air runoff. Relief valves are provided to purge GN_2
prior to propellant tanking or in the event of a purge supply failure.

(3) <u>LH_2/LO_2 Tank Intermediate Bulkhead</u>. The twin-skin vacuum bulkhead
separating the two tanks has been employed successfully on all Centaur
vehicles. As shown in Figure 3-10, the assembly contains a fiberglass
mat insulation that is maintained in compression by the spring ring
bulkhead and by LO_2 tank pressure. Cryopumping of the intervening
volume folliowng LH_2 tanking yields a low conductance of less than
0.045 Btu/hr-ft^2-R. A typical measured heating rate through the
intermediate bulkhead is 2350 Btu/hr. This corresponds to an effective conductance of only 0.0393 Btu/hr-ft^2-R.

(4) <u>LO$_2$ Tank Aft Bulkhead Radiation Shield</u>. This shielding system
consists of three double aluminized, scrim-reinforced Kapton radiation
shields, and a fourth with a teflon outer surface. LO$_2$ tank aft
bulkhead heat flux through the shielding system was measured at 1.0 to
2.0 Btu/hr-ft^2 on TC-5.

(5) <u>LO$_2$ Tank Sidewall Radiation Shield</u>. This shielding system is simply
an extension of the LH$_2$ radiation shield assembly although lighter
weight. A helium-purged foam blanket is not required for prelaunch
insulation of the LO$_2$ tank sidewall, and is thus omitted in this
system, and the lighter Nomex scrim-reinforced laminates are used.

3.3.2.2 <u>Centaur Integrated Support System (CISS)</u>. The CISS consists of a
Centaur support structure, a deployment adapter, and the associated CISS
electronics, fluid, and mechanical systems.

a. <u>Centaur Support Structure (CSS)</u>. The CSS supports the Centaur vehicle and
cantilevered payload within the Orbiter payload bay throughout the STS/
Centaur mission and during ground handling operations. The CSS transfers
all Centaur axial loads to the Orbiter through the pins at X_o = 1171.27
and shares the transfer of side loads with the forward adapter through the
keel pin at X_o = 1226.33. Vertical loads are reacted partly at the
forward adapter and the CSS pins at X_o = 1171.27 and 1273.53. The
trunnion pins at X_o = 1237.5 allow the deployment adapter and Centaur to
rotate before separation (see Figure 3-19).

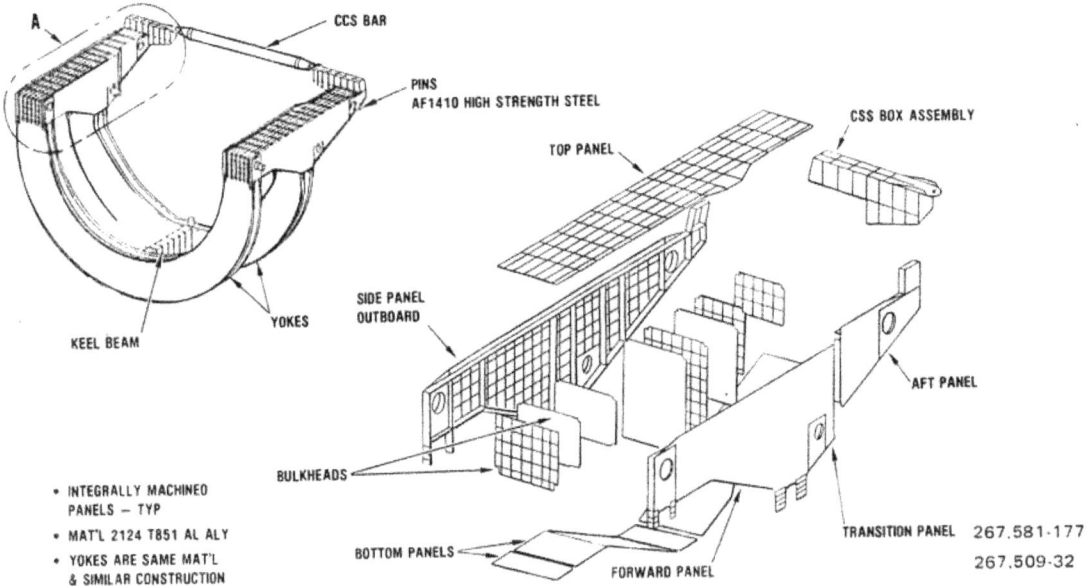

Figure 3-19. Centaur Support Structure (CSS)

The CSS structure is aluminum with the exception of the interface pins which are high strength steel. Basic construction is conventional airframe box-beam type with bulkheads and skins. Major elements of the CSS are two side beams, two circular beams, and a 6-inch-diameter tubular strut that spans the payload bay and connects the side beams. The two circular beams consist of integrally machined panels which make up the fore and aft faces of each yoke. Separating these panels is a series of machined internal ribs. Closing the yoke assemblies on the inboard and outboard surfaces are 2024 Al skins. A riveted keel beam connects the two circular beams. The two side beams consist of large, integrally machined panels built around nine integrally machined bulkheads (Figures 3-20 through 3-22). The interface pins (figure 3-23) are hollow for maximum strength-to-weight ratio.

Wherever possible, only high-stress corrosion-resistant materials are used. All structural components are designed to provide ultimate factors of safety equal to or greater than 1.40 for all mission design phases. Components are designed to the limit load factor specified in ICD-2-1F001. The CSS has been analyzed in considerable depth with detailed finite-element models. Optimization techniques developed by GDC during Orbiter midfuselage design have been used extensively in the design of the structure.

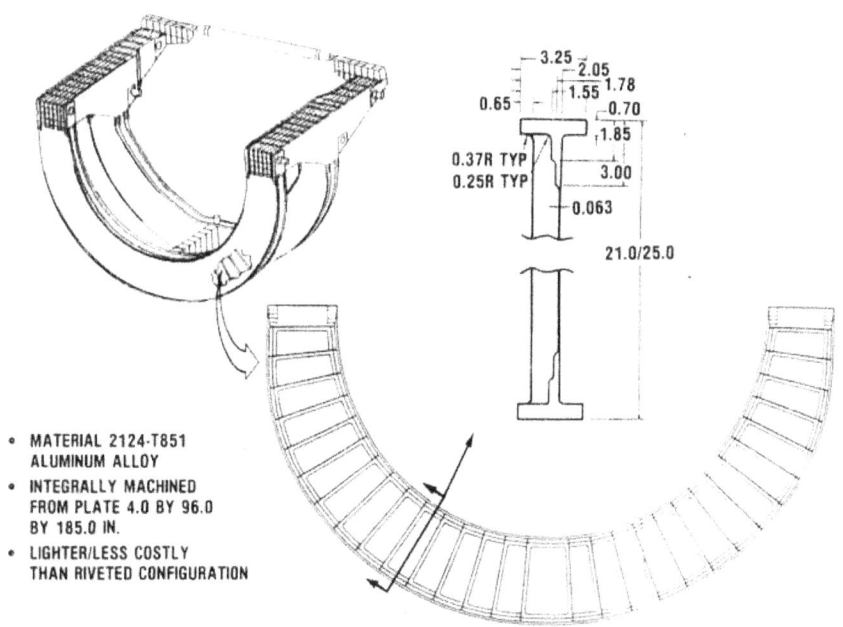

267.509-33

Figure 3-20. CSS Circular Beam Panels

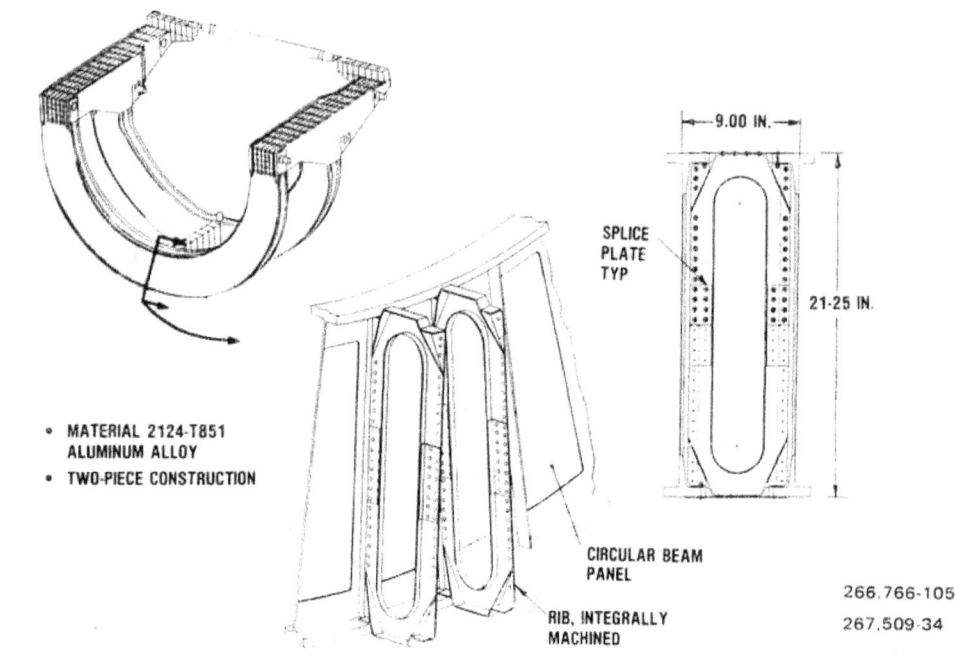

Figure 3-21.   CSS Circular Beam Ribs

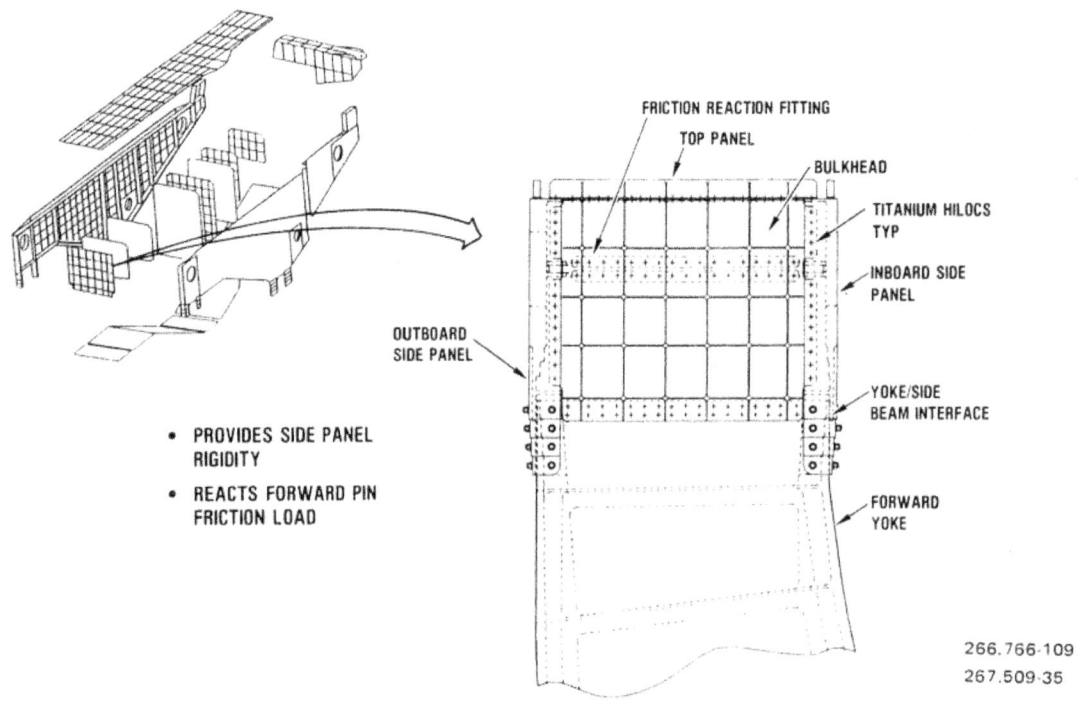

Figure 3-22.   CSS Bulkhead Installation

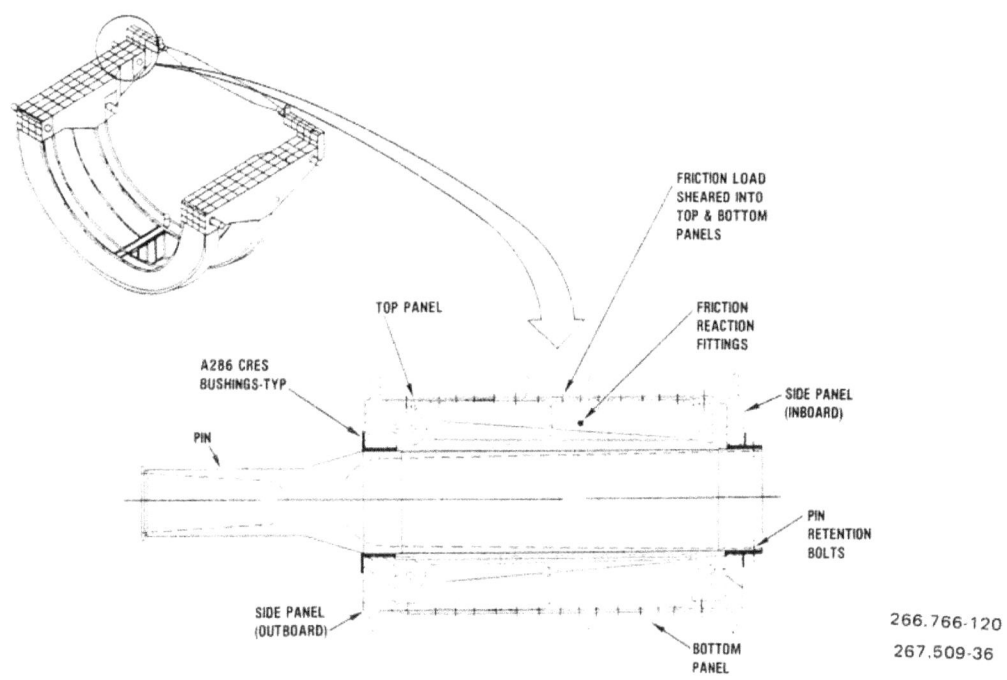

Figure 3-23. CSS Steel Interface Pins

b. <u>Centaur Deployment Adapter</u>. The Centaur G-prime deployment adapter
(Figure 3-24) interfaces with Centaur through the Super*Zip separation
ring. The adapter effectively transfers Centaur loads to the CSS during
flight. The basic structure of the 10-foot-diameter, 44-inch-high adapter
is conventional aluminum skin-stringer frame. It has four rings; the
three forward rings are machined and the aft ring is a web, cap, and
stiffener assembly. The skin thickness is variable with stringers spaced
at six degrees.

The adapter rotates about the CSS trunnion support pins during deployment.
The machined aluminum fittings that interface with these pins via bearings
also transfer Centaur vertical and axial loads during flight. A semicir-
cular channel on the adapter interfaces with the aft lower keel guide pin
on the CSS to react lateral loads during flight and deployment.

The deployment adapter supports the rotation mechanisms, spring thrust
system, two fluid umbilical panels, the electrical umbilical panels,
Centaur engine support structure, numerous valve panels, and avionic
support mounting shelf.

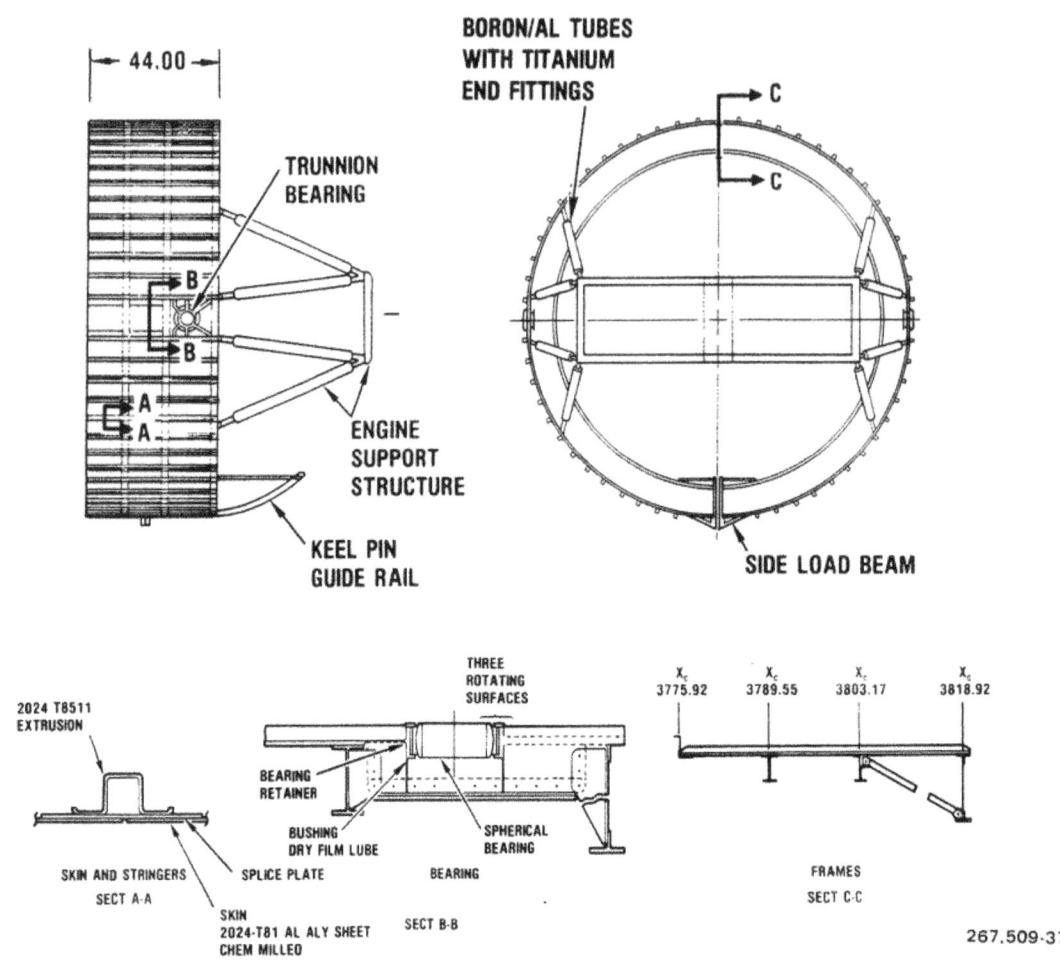

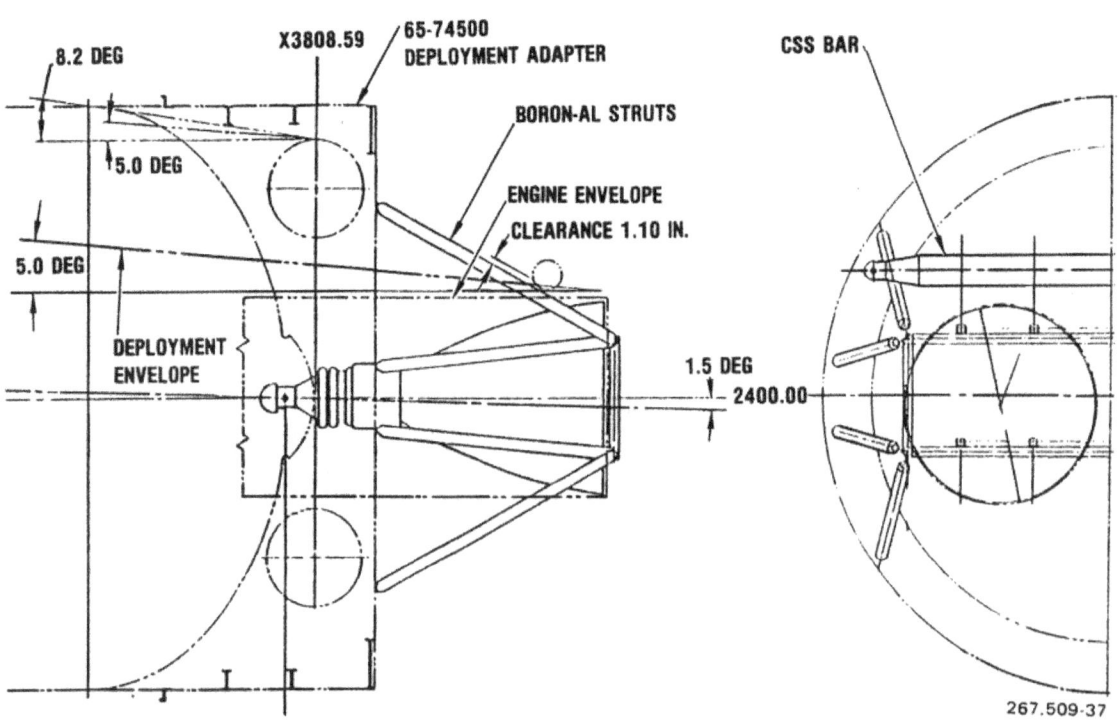

Figure 3-24. Deployment Adapter and Engine Support Structure

c. Centaur Engine Support Structure. The engine support structure, also shown in Figure 3-24, is a lightweight boron/aluminum tube truss arrangement that supports the Centaur engines cantilevered from the aft bulkhead, and aids in load redistribution by the deployment adapter. The truss mounts on the aft end of the deployment adapter. The aft end of the truss is an extruded aluminum support frame that provides for attachment of engine support brackets. These brackets slip inside the engine bells and support them through a nonmetallic bearing ring. Centaur forward motion during deployment separates this passive support system.

d. Centaur Spring Thrust System. The Centaur G-prime spring thrust system consists of twelve spring-loaded pushrods on the deployment adapter that bear against reaction gussets on the aft adapter (Figure 3-25). The springs provide the thrust to separate Centaur from the Orbiter at a Centaur-to-Orbiter separation velocity in excess of one foot per second. The springs are designed to safely separate the Centaur from the Orbiter under worst case failure of two springs. The guides shown in Figure 3-26 guarantee at least one inch clearance under worst case conditions while separating from the CISS. The guides also guarantee clearance of the Centaur above the crew cabin as shown in Figure 3-27.

e. Centaur Mounting Provisions. Several subsystems have mounting provisions on the deployment adapter. The avionics packages are conveniently mounted on the accessible upper side of the deployment adapter (Figure 3-28). Other systems mounted on the deployment adapter and aft adapter are the fluids (lines and valves), electrical (wiring), insulation (radiation shields), and air conditioning (lines) systems.

The CSS provides mounting provisions for 20 helium bottles. The bottle support structure consists of a bipod strut arrangement stabilizing one side of each bottle. On the opposite side a tripod strut system resists movement in three directions and also prevents rotation by providing a keyed interface (Figure 3-29). The CSS also supports fluid lines, the deployment drive system, and electrical packages.

f. CISS Air-Conditioning Duct (Aft End). A 6-inch-diameter duct extends forward from the CSS (Figure 3-30) and butts with the existing Orbiter interface at $X_o = 1127.2$ through a compressed bellows. The gas is routed through the CSS keel beam and interfaces with the deployment adapter through a riseoff bellows disconnect.

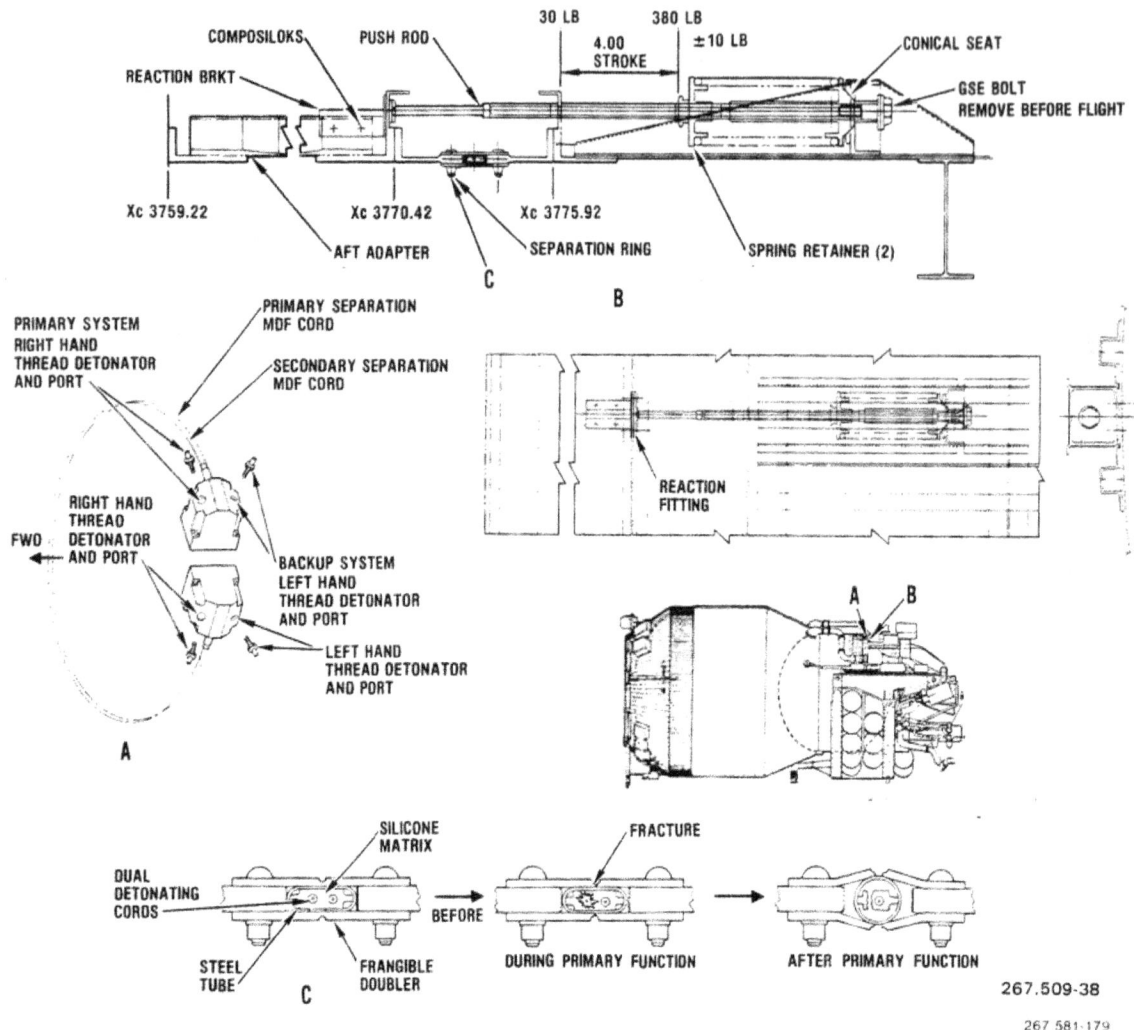

Figure 3-25. Separation System

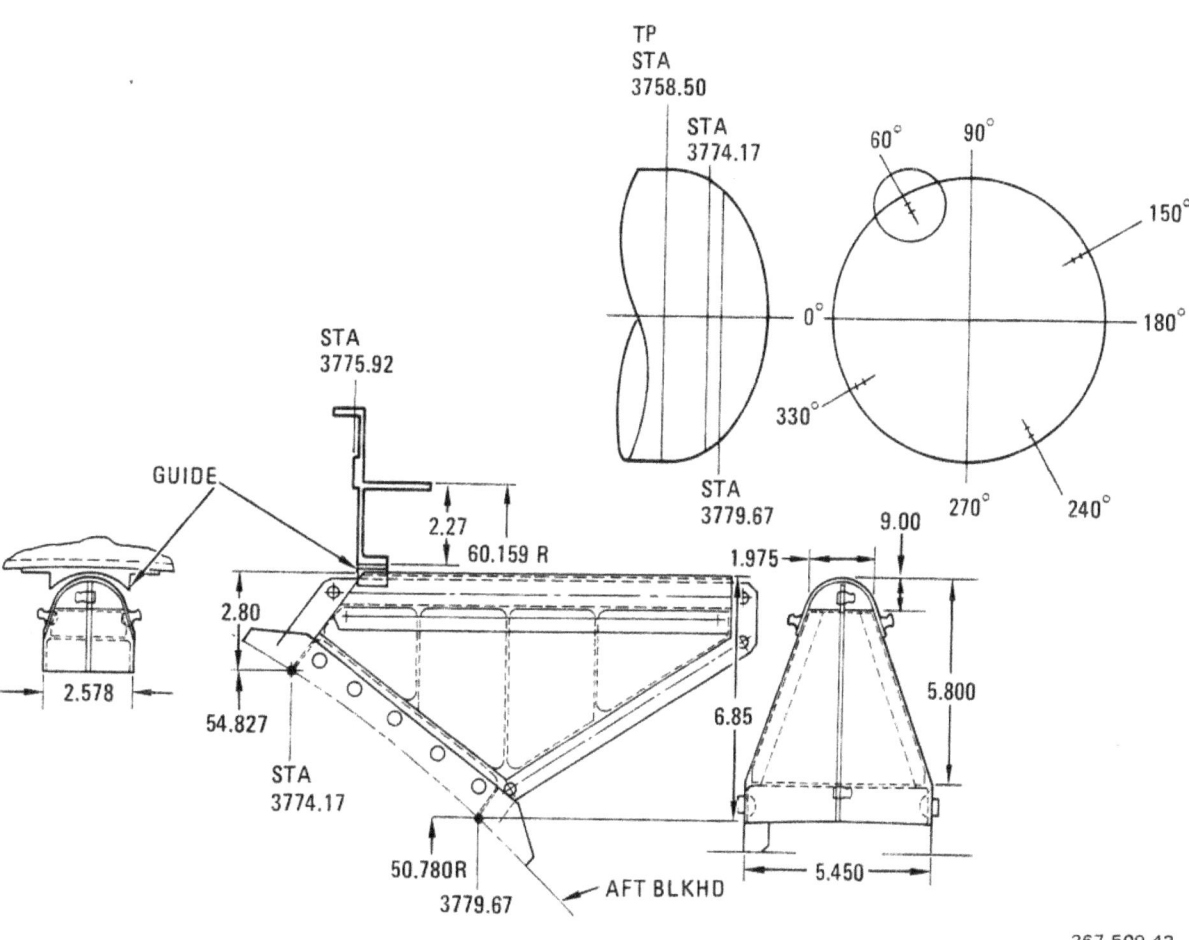

Figure 3-26. Centaur CISS Separation Guides

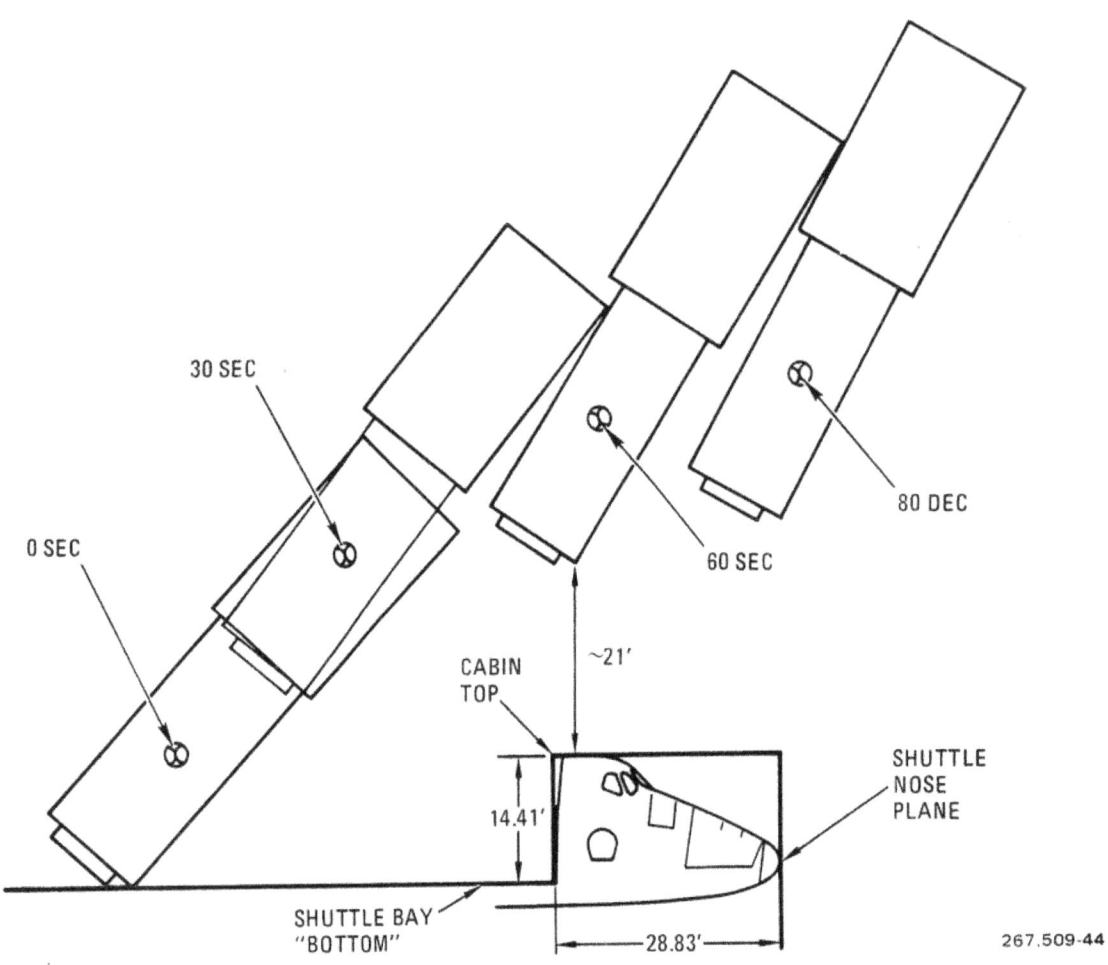

Figure 3-27. Galileo Pitch-Up Worst Case Cabin Clearance

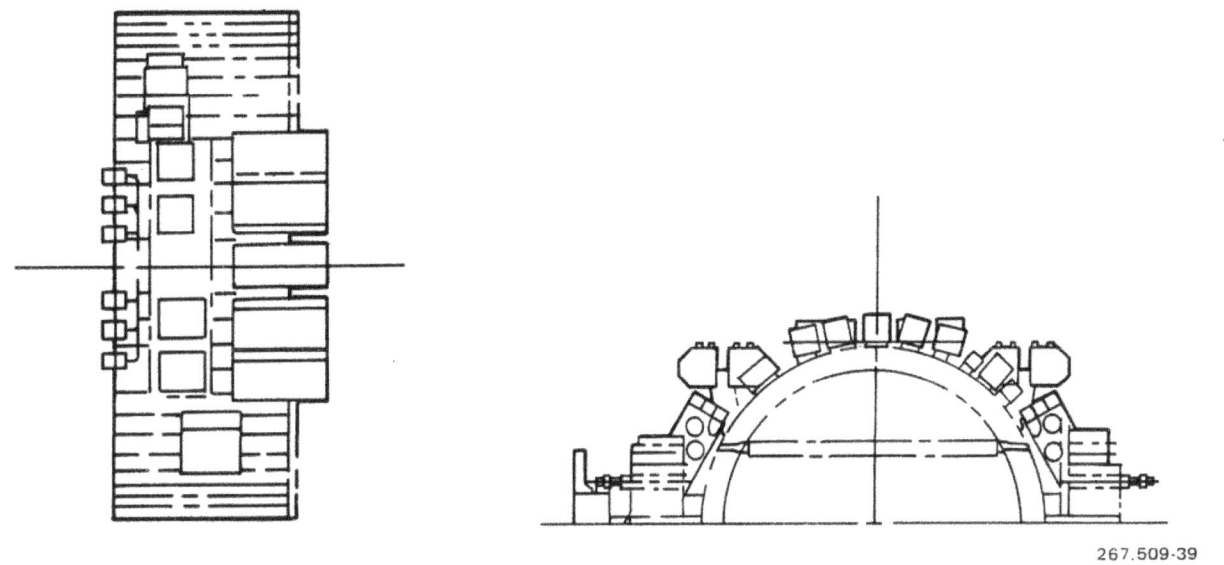

267.509-39

Figure 3-28. Deployment Adapter Shelf

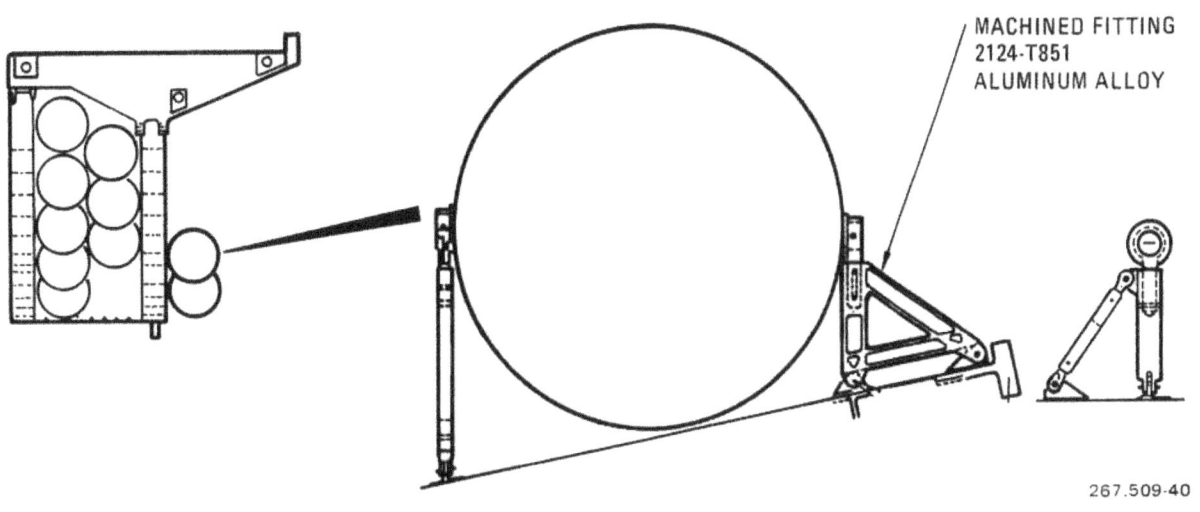

MACHINED FITTING
2124-T851
ALUMINUM ALLOY

267.509-40

Figure 3-29. CISS Helium Bottle Support Structure

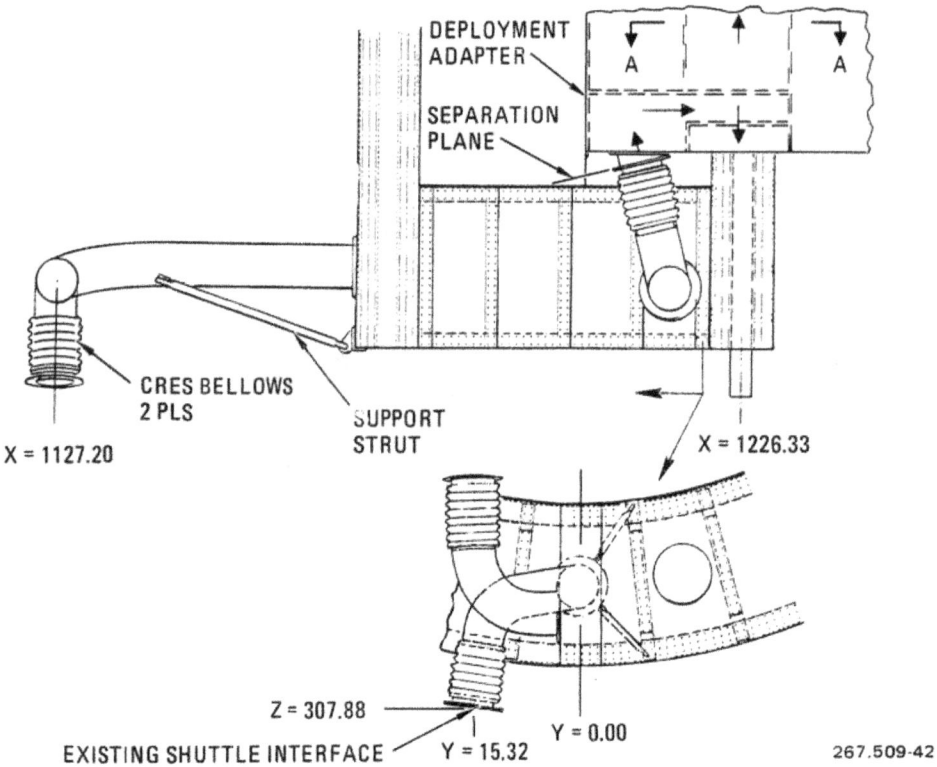

CRES BELLOWS
2 PLS

X = 1127.20

DEPLOYMENT
ADAPTER

SEPARATION
PLANE

A   A

SUPPORT
STRUT

X = 1226.33

Z = 307.88

EXISTING SHUTTLE INTERFACE   Y = 15.32   Y = 0.00

267.509-42

Figure 3-30.   CISS Air-Conditioning Duct Components

g. <u>Orbiter Interfaces with Centaur G-Prime</u>.   Centaur and CSS will be
contained within the 15-foot-diameter payload thermal and dynamic envelope
throughout all operational phases.   CSS attachment fittings will extend
beyond this envelope at five locations, and Centaur attachment fittings
will extend beyond this envelope at three locations to mate with the
Orbiter attachment fittings, which will be outside of the envelope.

(1) <u>CSS Interface</u>. Two aft retention pins at the Orbiter sill longeron at
X_0 = 1273.53 share the aft vertical load. Two forward retention
pins at X_0 = 1171.27 (Figure 3-31) share the vertical load with the
aft pins, and react all axial loads from the Centaur with its payload.
The Orbiter latches for these four pins are of the passive type. The
keel pin at X_0 = 1226.33 will react a portion of the lateral load.

GN$_2$ conditioning ducts with bellowed ends to accommodate relative
motion interface with the Orbiter standard spigots as shown in Figure
3-32.

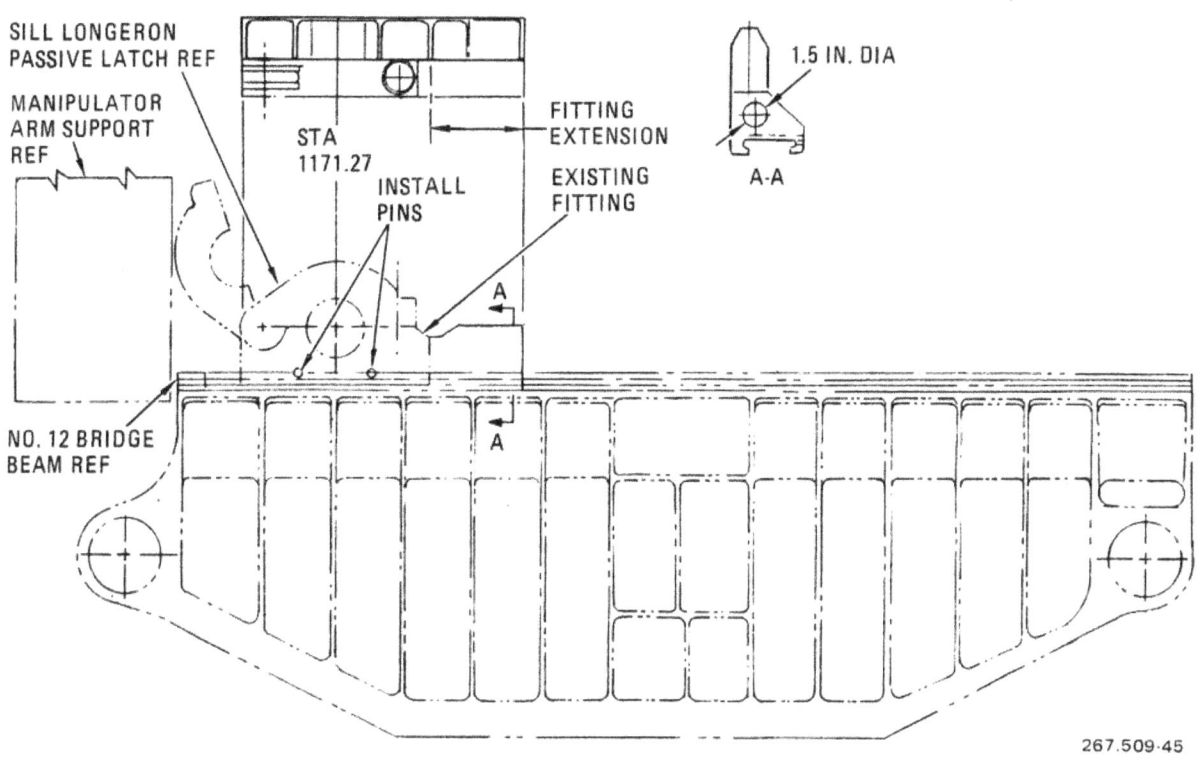

Figure 3-31. CSS Adjustable Primary Longeron Fitting

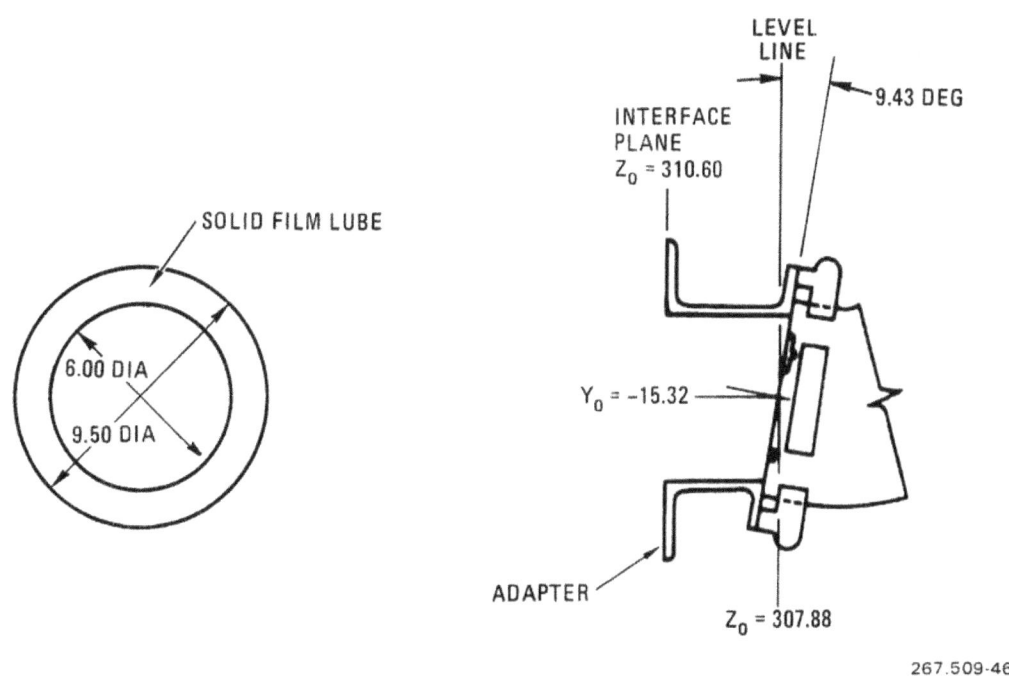

Figure 3-32. CISS GN$_2$ Conditioning Interfaces

(2) <u>Forward Adapter Interface.</u> The two retention pins on the Centaur forward adapter at the Orbiter sill longeron share the vertical load with the CSS pins. They are located at X_O = 950.50 and interface with Orbiter active attach fittings for deployable payloads. The bridge beams for these latches are modified slightly to add latch stops for Centaur/spacecraft vehicle support should it be necessary to depressurize the tank (Figure 3-33). The keel pin on the Centaur equipment module at X_O = 950.50 shares the lateral load with the CSS keel pin. This pin interfaces with an active keel fitting on the Orbiter. The bridge beam for this latch is modified to add a latch stop to act with the longeron latch stops in event of tank depressurization (Figure 3-34).

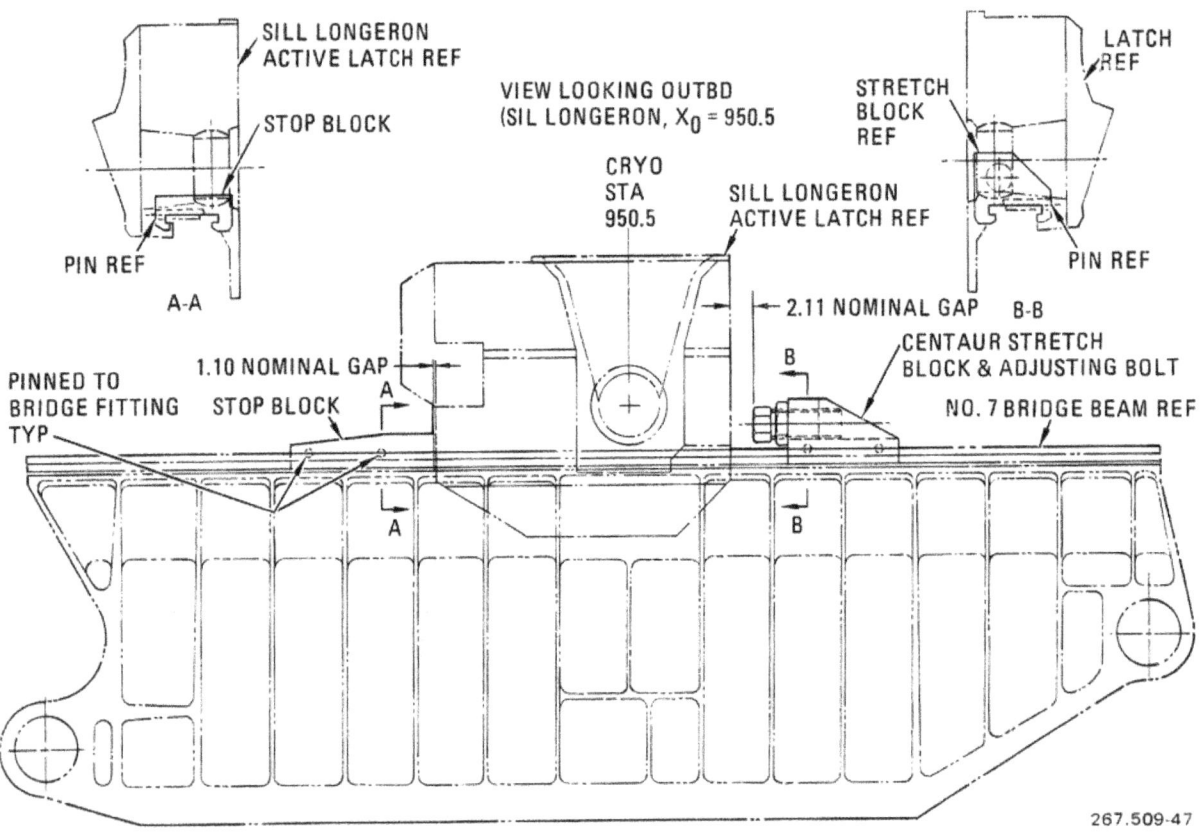

Figure 3-33. Active Forward Sill Latch Stop and Stretch Fittings

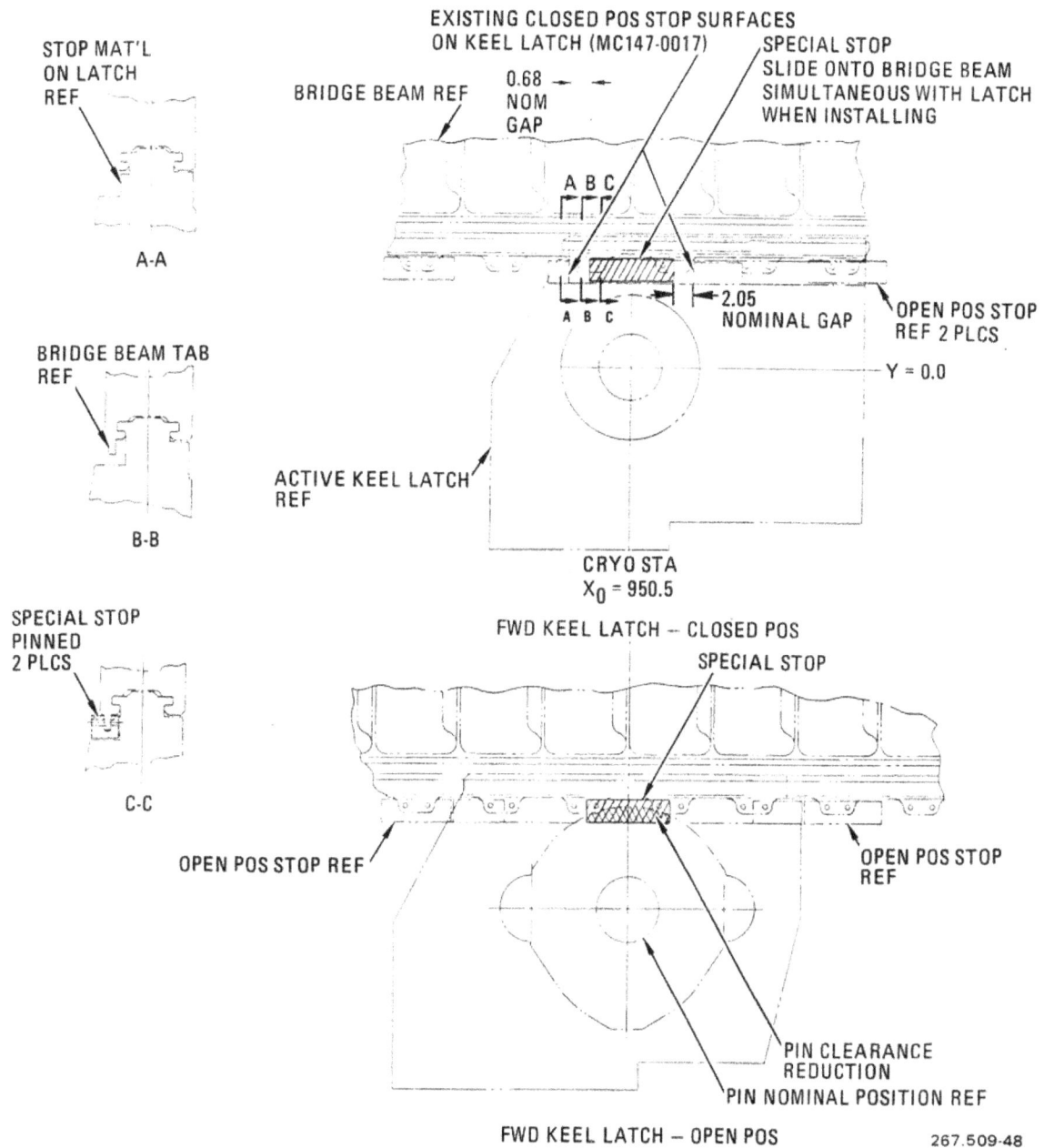

Figure 3-34. Active Forward Keel Latch Stop

3.3.3 <u>MECHANISMS</u>. The Centaur G-prime mechanisms (Figure 3-12) consist of identical primary and backup deployment adapter rotation systems. These systems meet the intent of the two-failure tolerant requirements without the use of crew-trained EVA. Both the primary and backup systems are tolerant to single failures of the drive motors and clutch motors and either can rotate the deployment adapter under maximum expected loading conditions.

The deployment adapter primary rotation system shown in Figure 3-35 consists
of a deployment adapter-mounted power drive unit, a crank clutch unit, a
crank, and a link connecting the crank to the CSS. Counterclockwise rotation
of the crank causes the deployment adapter to rotate clockwise about the
trunnion and pivot the Centaur/spacecraft out of the payload bay into the
deployment position.

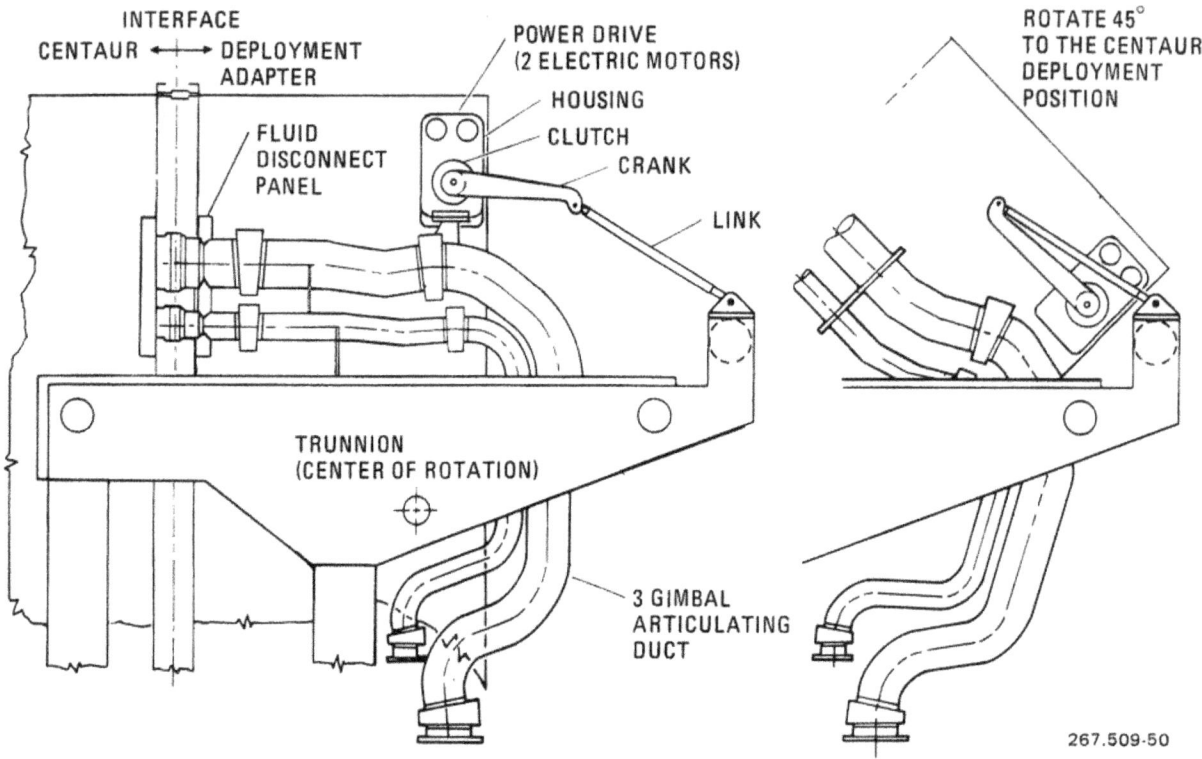

Figure 3-35. Deployment Adapter Rotation System

To start rotation, the primary rotation system clutch is engaged, the Orbiter
forward keel and sill latches are released, and electrical power from two-
failure tolerant Orbiter 400 Hz, 115-volt, three-phase buses is applied to one
of the two drive unit motors (see Figure 3-36). The brake releases, the motor
clutch engages, and torque is transmitted and multiplied via the differential
and planetary gear train through the clutch to the crank. This design
provides single-failure tolerant motor operation. If the first motor should
fail to operate, its brake keeps it from turning and permits the second motor
to drive the load at the same speed. The double-gear reduction ratio, which
occurs when one side of the differential is locked, doubles the output torque
of the single driving motor.

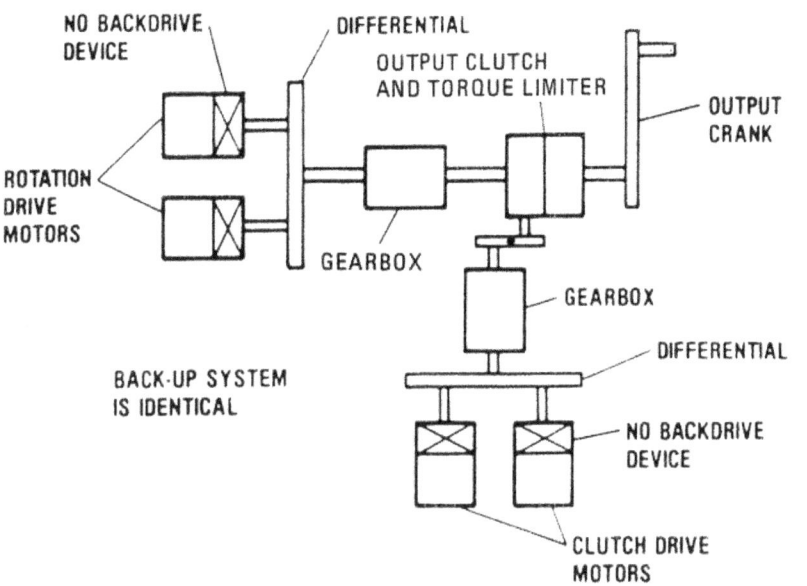

Figure 3-36. Rotation System Functional Block Diagram

With one motor operating, the deployment adapter rotates 45 degrees in approximately five minutes. Once rotated, software will measure the change in crank angle over a period of time. If no change occurs, the controllers will assume full rotation and signal the controller units to remove power from the drive unit, causing the brake to set and the motors to stop. The deployment adapter is held in the rotated position by the deenergized drive motor brake and requires no separate latch.

Following Centaur deployment, the payload bay doors may be closed with the deployment adapter in any position between 0 and 45 degrees.

Should deployment be aborted, the Centaur is rotated back to the 0-degree position by reversing the drive unit. The redundant crank position transducers signals (not shown) are used to stop the rotation-system power drive unit motors. The Orbiter latches are then relatched and propellants are dumped for reentry and landing.

The backup rotation system is identical to the primary. The backup system is mounted on the opposite side of the deployment adapter.

During normal primary system operation, the crank of the backup system is disconnected from its drive unit by the disengaged clutch and the linkage merely follows the rotation. Should both primary motors fail, the crankshaft clutch on the backup system is automatically engaged and then the primary clutch disengaged. Rotation then continues in either direction using the backup power drive. Manual disengagement capability of the link for contingency only will be provided, but crew-trained EVA is not required.

3.3.4 FLUID SYSTEMS. The Centaur G-prime vehicle fluid systems consist of hydraulic, pneumatic, main propulsion/propellant supply, hydrazine reaction control, vent, and fill/dump systems.

3.3.4.1 Centaur Vehicle Fluid Systems. Refer to the Fluid Systems Schematic Legend following the list of acronyms to interpret the fluid systems schematics shown in this section.

a. Hydraulic System. The Centaur hydraulics has identical independent systems for each main engine as shown in Figure 3-37. Each system supplies the hydraulic power for the two closed-loop, servo-controlled engine gimbal actuators that provide thrust vector control. Major assemblies of each hydraulic system are the power package assembly, containing an engine-driven main pump and an electric motor-driven recirculation pump; and two servo cylinder assemblies that react in accordance with guidance and flight control system commands.

The hydraulic system is inactive in the payload bay except for operation of the low-pressure recirculation motor during liftoff and abort landing and intermittent operation on-orbit for thermal conditioning. Hydraulic system fluid is MIL-H-5606. A helium purge of the electric recirculation pump motors provides explosion-proofing. Recirculating pump operating pressure is 110 psig.

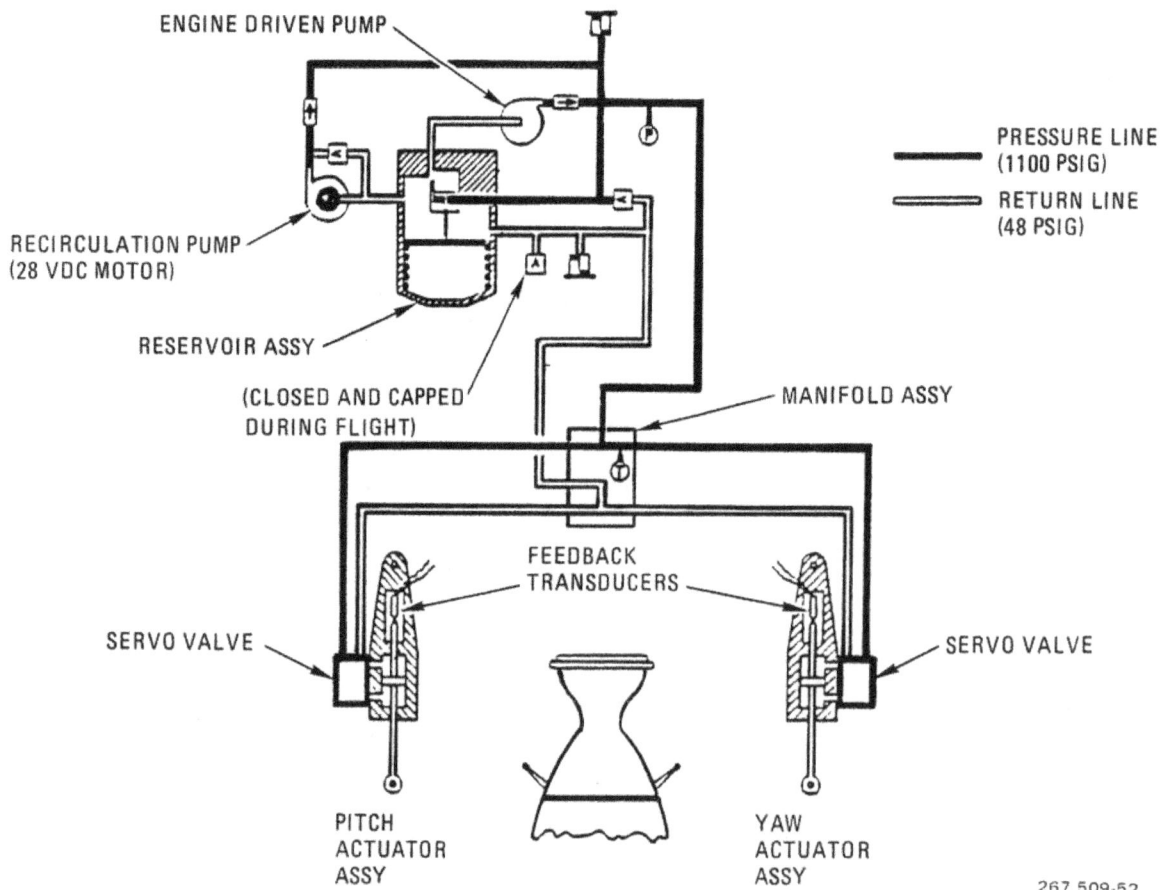

Figure 3-37. Centaur Hydraulic System

b. Pneumatic System. The Centaur pneumatics systems consist of the propellant tank pressurization system, pneumatic actuation valve control
system, purge systems, helium supply, and the intermediate bulkhead relief
system (Foldout FO-1, foldouts are placed in the back of the book). These
systems are interconnected and function as a single system, except for the
bulkhead relief system which is completely autonomous. Each is discussed
in the following section.

(1) Pressurization System (Figure 3-38). The Centaur pressurization
system consists of solenoid valves, tubing, and associated components
for pressurizing the propellant tanks, using either the 12 valves
located on Centaur or via CISS valves. During Centaur free flight, a
parallel set of two series solenoids for each tank are cycled upon
command from the Centaur digital computer unit (DCU) to provide helium
pressurant control before engine starts and also for the LO_2 tank
during engine burns.

An additional parallel set of two series solenoid valves and bypass
orifice are used to control the LH_2 tank pressure during engine
burns by using gaseous hydrogen bleed from the main engines. The DCU
determines the respective propellant tank pressures by monitoring
outputs from three of four redundant transducers in each tank.
Preprogrammed logic defines the desired pressure levels and sequences
throughout the mission. These propellant tank pressurization
techniques have been demonstrated through testing, and have been
thoroughly analyzed for all configurations of the Centaur D-1A.

Before deploying the Centaur from the Orbiter, the quad set of helium
pressurization valves for each propellant tank is controlled by the
five control units located on the CISS, using outputs from five
redundant pressure transducers in each tank. This provides the third
method of pressurizing the propellant tanks at high flow rates (two
methods are provided by the CISS system) to ensure a two-failure
tolerant pressurization system for return to launch site propellant
dump and preliftoff propellant tank pressurization.

(2) Pneumatic Actuation Valve Control System (Figure 3-39). The pneumatic
valve actuation system consists of two parallel sets of tubing, check
valves, and solenoid valves that provide a nominal 450 psig supply
pressure for actuation of the two propellant tank fill/dump valves,
the vent shutoff valves in each tank, and the LO_2 JPM, as well as
the four LH_2 tank zero-g vent system valves, the pyro-isolated
propellant isolation prevalves, and the LH_2 TVS valves. The
pneumatic actuation valve control system also provides pressure for
the propellant tank sense line purges and the IRU, JPM, and NRM
purges. One branch passes through a disconnect in the Centaur/CISS
oxidizer umbilical panel. The other parallel branch passes through a
disconnect in the Centaur/CISS fuel umbilical panel.

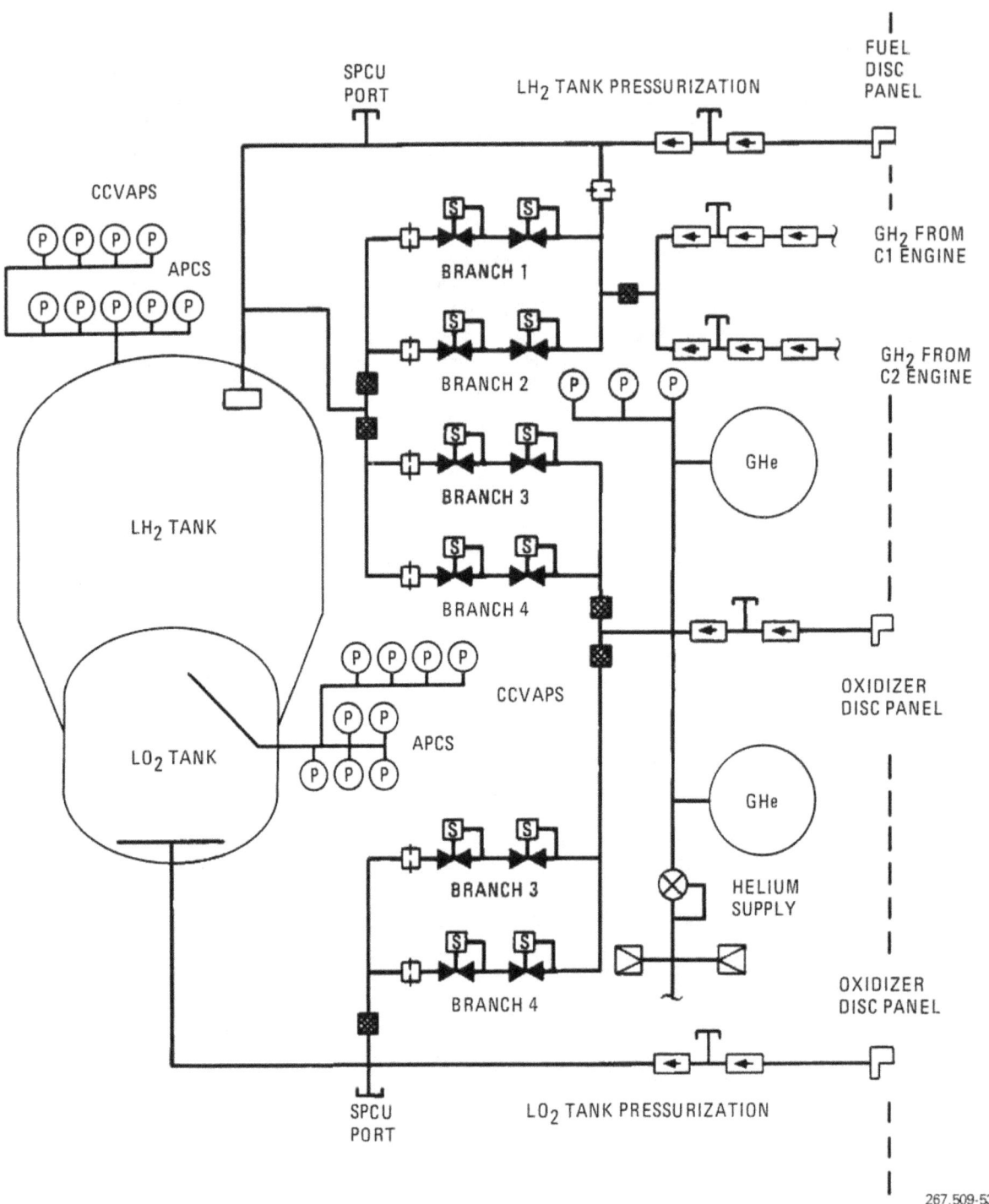

Figure 3-38.   Centaur Pressurization System

267.509-53

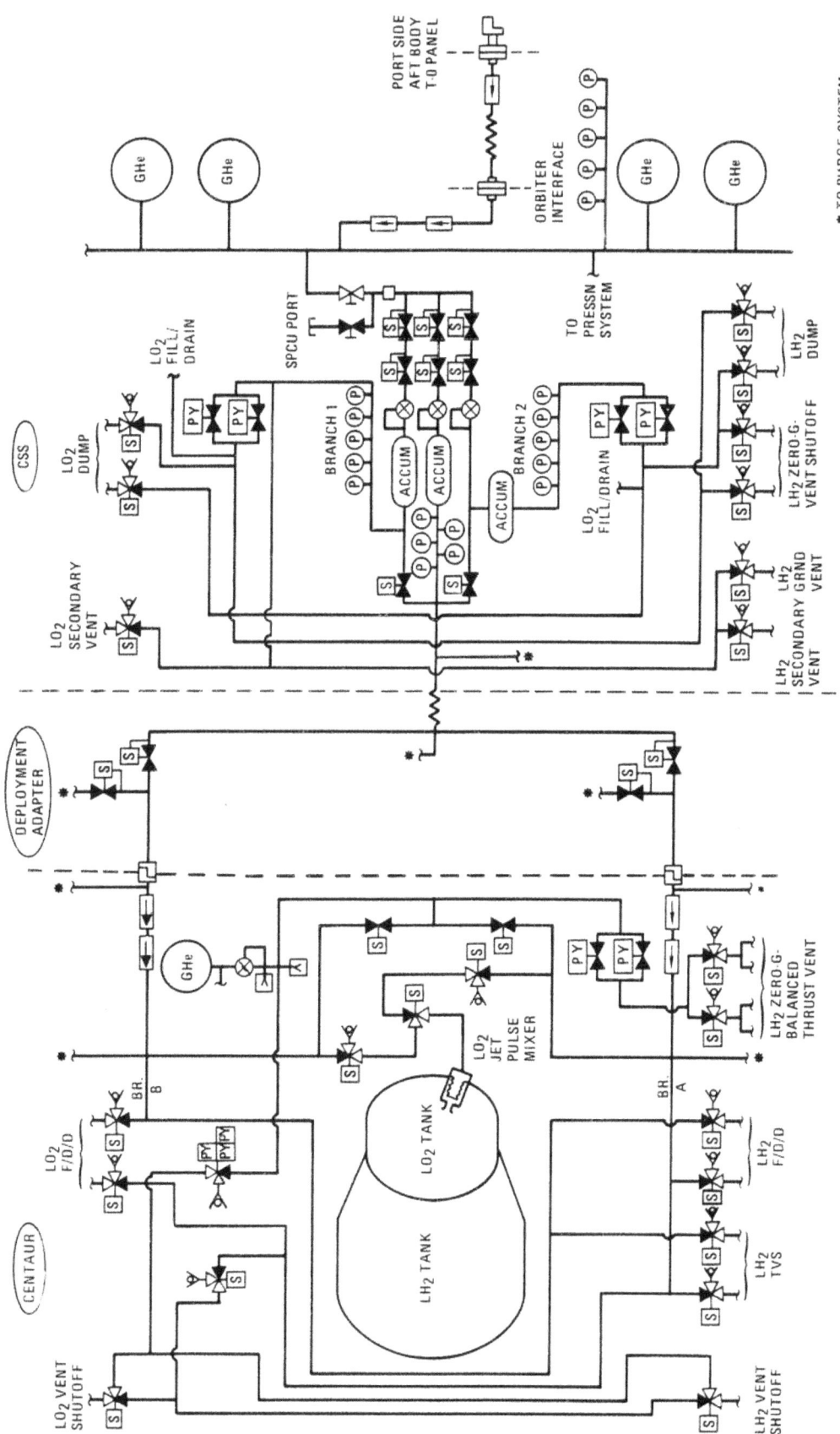

Figure 3-39.   Pneumatic Actuation Valve Control System (PAVCS)

267.509.64

Actuation pressure is supplied from the two-failure tolerant pressure regulation system on the CISS. One branch supplies actuation pressure to one valve in each parallel set. The other line supplies the mating valve in each parallel set. The vent shutoff valves are supplied from one branch and the Centaur helium supply. Two solenoid valves provide the capability to supply either or both parallel branches from the 450-psig pressure regulation system on Centaur. Check valves provide Centaur pneumatic branch isolation after Centaur deployment.

(3) Purge System (Figure 3-40). The purge system consists of solenoid valves, flow control orifices, tubing, and components to direct helium purges to various locations on the vehicle at various times during the mission. One line routes from a disconnect in the Centaur/CISS fuel disconnect panel to supply helium purge to the LH$_2$ tank insulation system.

The insulation system is purged with helium gas introduced at ambient temperature and at up to 160 pounds per hour before launch. As shown in Figures 3-17 and 3-18, a Kapton/glasscloth/Kapton laminate containment membrane maintains positive internal helium pressure during prelaunch operations. Three ΔP transducers sense the insulation system purge pressure for proper control of purge supply via valves located on the CISS. Two relief valves preclude insulation system overpressure if the purge supply fails open.

There is no insulation system purge flow requirement during boost phase. A vent door is opened to enable rapid evacuation of the helium-purge blanket to prevent blanket overpressurization and achieve effective inflight radiation shielding of the LH$_2$ tank sidewall. In the event of an inflight abort, the vent door will close to permit blanket repressurization during the reentry sequence. The three ΔP transducers will be employed to maintain blanket pressure control during this period.

A continuous purge is supplied to the LO$_2$ and LH$_2$ tank pressure transducer sensing lines. A single solenoid valve provides helium from the Centaur helium supply for purges going to the LO$_2$ tank vent standpipe and the LO$_2$ tank pressurization line. A purge is supplied through a common manifold to the two hydraulic system recirculation pump electric motors, the JPM and the IRU from the pneumatic valve actuation system supply coming from Centaur/CISS.

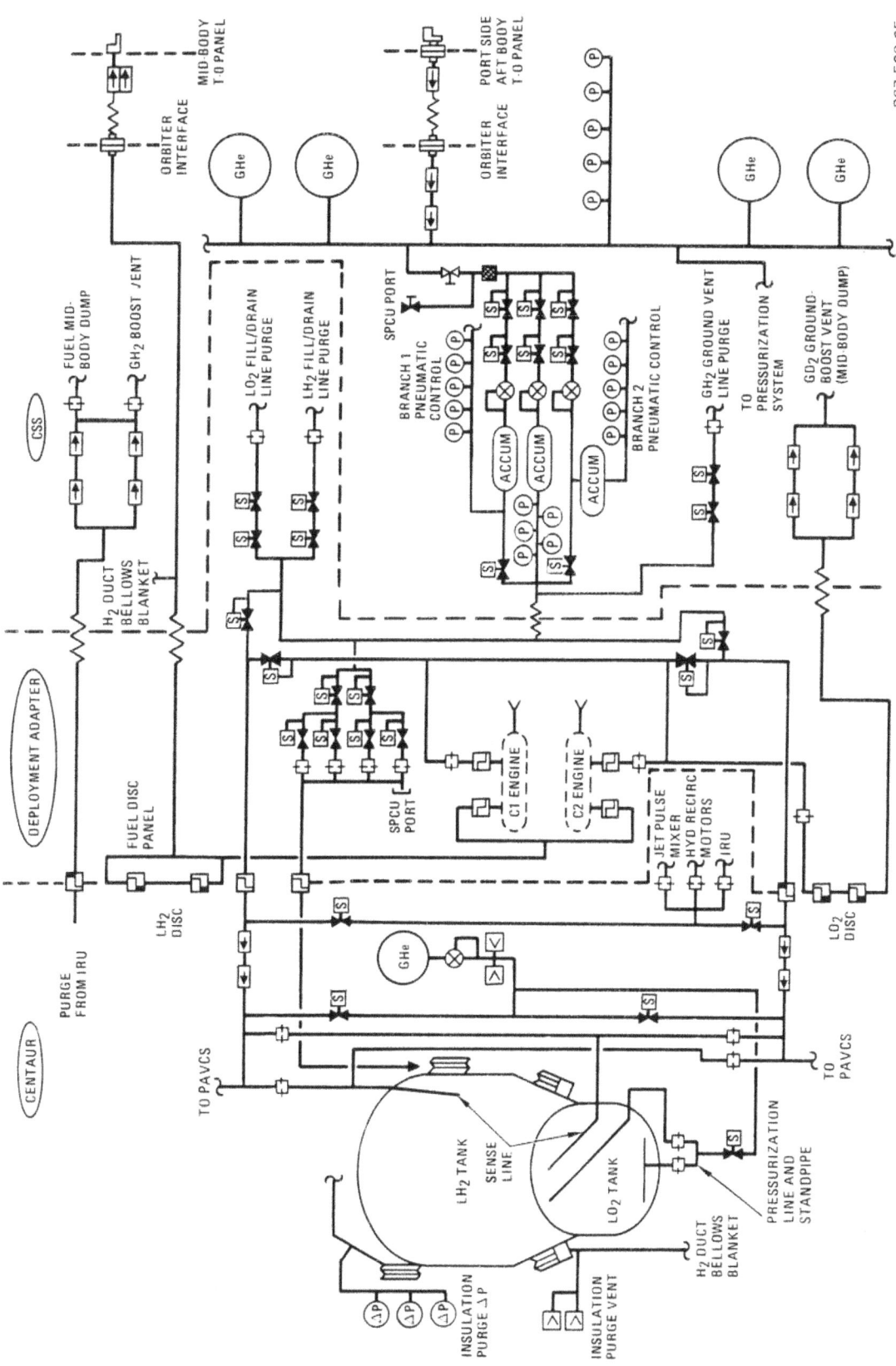

Figure 3-40. Purge System

(4) Centaur Helium Supply System (Figure 3-41). The helium supply system consists of the two Kevlar overwrapped helium storage spheres and a charge line connected to a disconnect in the Centaur/CIUSS oxidizer umbilical panel through two series check valves. An analysis has determined the two 26-inch-diameter helium spheres will contain the helium mass required for the interplanetary mission. The helium charge is controlled by valves located on the CISS. The system also contains a regulator to provide the various purge systems, pneumatic actuated valves, and N_2H_4 bottle pressurization, with a nominal 450-psig supply. Two relief valves provide protection from overpressurization in the 450-psig portion of the system.

20 — PRESSURIZATION SYSTEM

23 — PNEUMATIC ACTUATED VALVE CONTROL SYSTEM

25 — ENGINE CONTROLS SYSTEM

26 — HYDRAZINE REACTION CONTROL SYSTEM

27 — PURGE SYSTEM

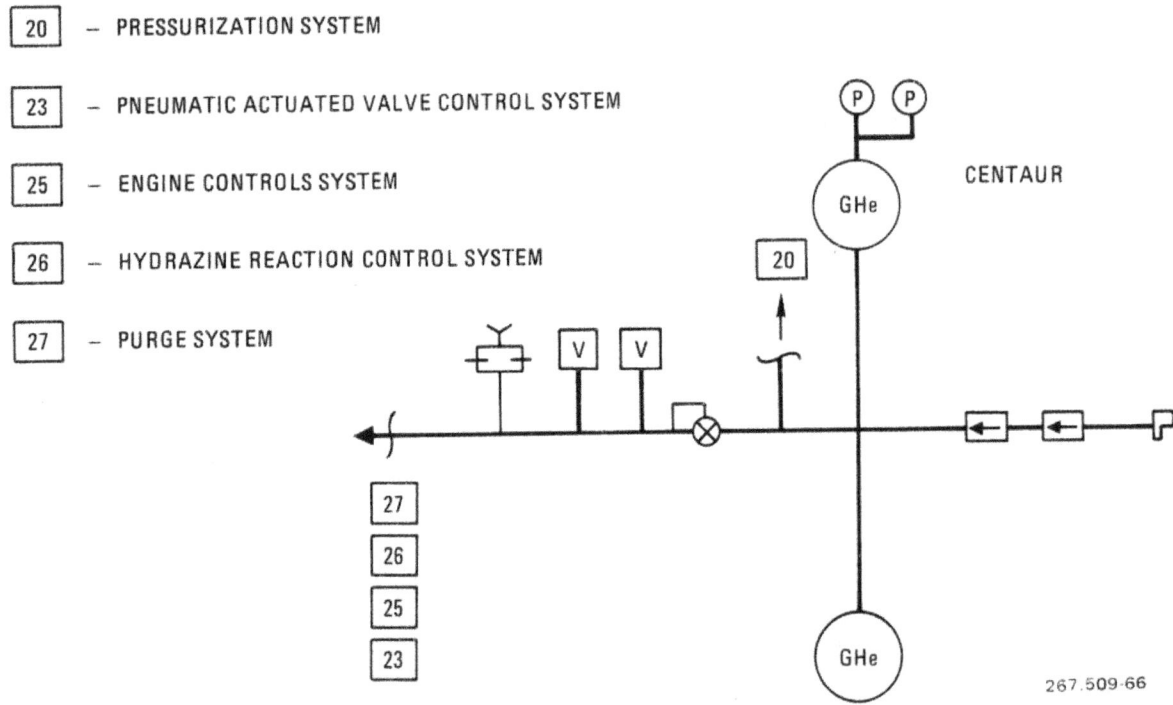

267.509-66

Figure 3-41. Centaur Helium Supply System

(5) Intermediate Bulkhead Relief System (Figure 3-42). The bulkhead relief system consists of tubing, pressure transducers (two measure vacuum), test port, and a single valve whose outlet is purged with GN_2 prior to liftoff. The capped test port and check valve allows the cavity to be evacuated and backfilled with GN_2. When the propellants are tanked, condensation of the GN_2 cryopumps the bulkhead cavity to a vacuum that provides the desired insulation between the LH_2 and LO_2 propellant tanks.

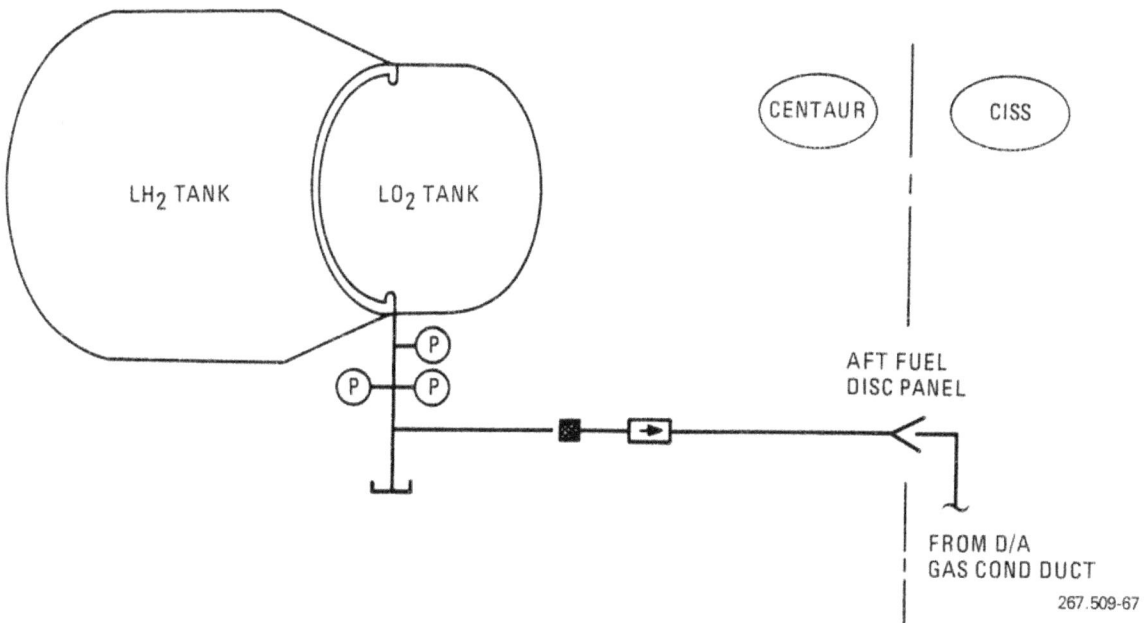

Figure 3-42.   Intermediate Bulkhead Relief System

c. <u>Main Propulsion/Propellant Supply</u>. The main propulsion system consists of
two Pratt & Whitney RL10A-3-3A engines rated at 16,500 lbf nominal thrust,
each operating at a 5:1 mixture ratio of oxidizer to fuel. A silver
throat results in a specific impulse of 446.4 seconds. The net positive
suction head (NPSH) required by the engine turbopumps is provided by
pressurizing the vehicle propellant tanks. Propellants are delivered to
the main engine turbopumps through feed ducts from the vehicle propellant
tanks. The feed ducts contain flex joints to accommodate engine gimbaling
and are overwrapped with a three-layer, double-aluminized Kapton radiation
shield.

Pneumatically actuated prevalves located at the propellant tank outlets
provide series redundant backup for the engine inlet shutoff valves
(Figure 3-43). A parallel set of pyro valves and solenoid valves upstream
of the pneumatic actuation control solenoid valves provides two-failure
tolerance against inadvertent opening of the engine inlet shutoff valves.
The pyro valves will be fired open after Centaur is deployed a safe
distance from the Orbiter.

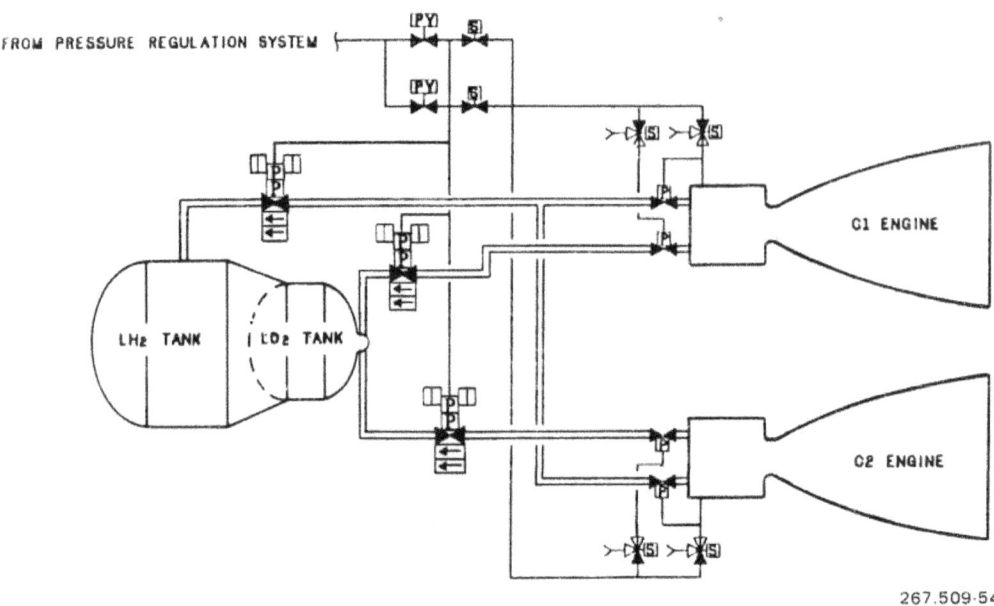

FROM PRESSURE REGULATION SYSTEM

C1 ENGINE

C2 ENGINE

LH₂ TANK   LO₂ TANK

267.509-54

Figure 3-43.   Propellant Feed and Main Engine Valve Actuation Control

d.   Reaction Control System.   The Centaur G-prime baseline system consists of twelve 6-lbf thrust units, a positive expulsion tank with 170-lbm hydrazine capacity, two sets of parallel pyro valves, one fill/drain and two pneumatic checkout valves, and both ground and inflight heaters (Figure 3-44).   All feed line joints, including those that interconnect components, are welded to provide a leakproof, contamination-free system. An optional, second hydrazine tank can be added when mission requirements dictate.

The fill/drain and pneumatic checkout valves (with redundant pressure sealing caps) are required to maintain a positive system GN₂ standby pressure, facilitate component functional checkout, and load the hydrazine tank.   A set of parallel pyro valves are used in the hydrazine tank inlet and outlet lines to provide positive isolation of the hydrazine tank.   The downstream set of parallel pyro valves and thruster series solenoid valves provide two-failure tolerance against inadvertent thruster operation.   The pyro valves will be fired open, pressurizing the system and allowing hydrazine flow to the thrusters after Centaur is deployed a safe distance from the Orbiter.   The arming mechanism is provided by the DUFTAS and is two-failure tolerant against inadvertent operation.

Prelaunch thermal control is provided by ground heaters on the hydrazine valve panel, tank, and supply line between the panel and tank.

Inflight thermal control of the system includes a ground energized heater and multilayered insulation over the tank shell and rocket engine modules (REM), periodic thruster warming firings, and flight heaters on all lines REMs, and hydrazine valve panel.   The reaction control system is tolerant of a thruster valve failure to close by series redundant solenoid valves on each thruster.

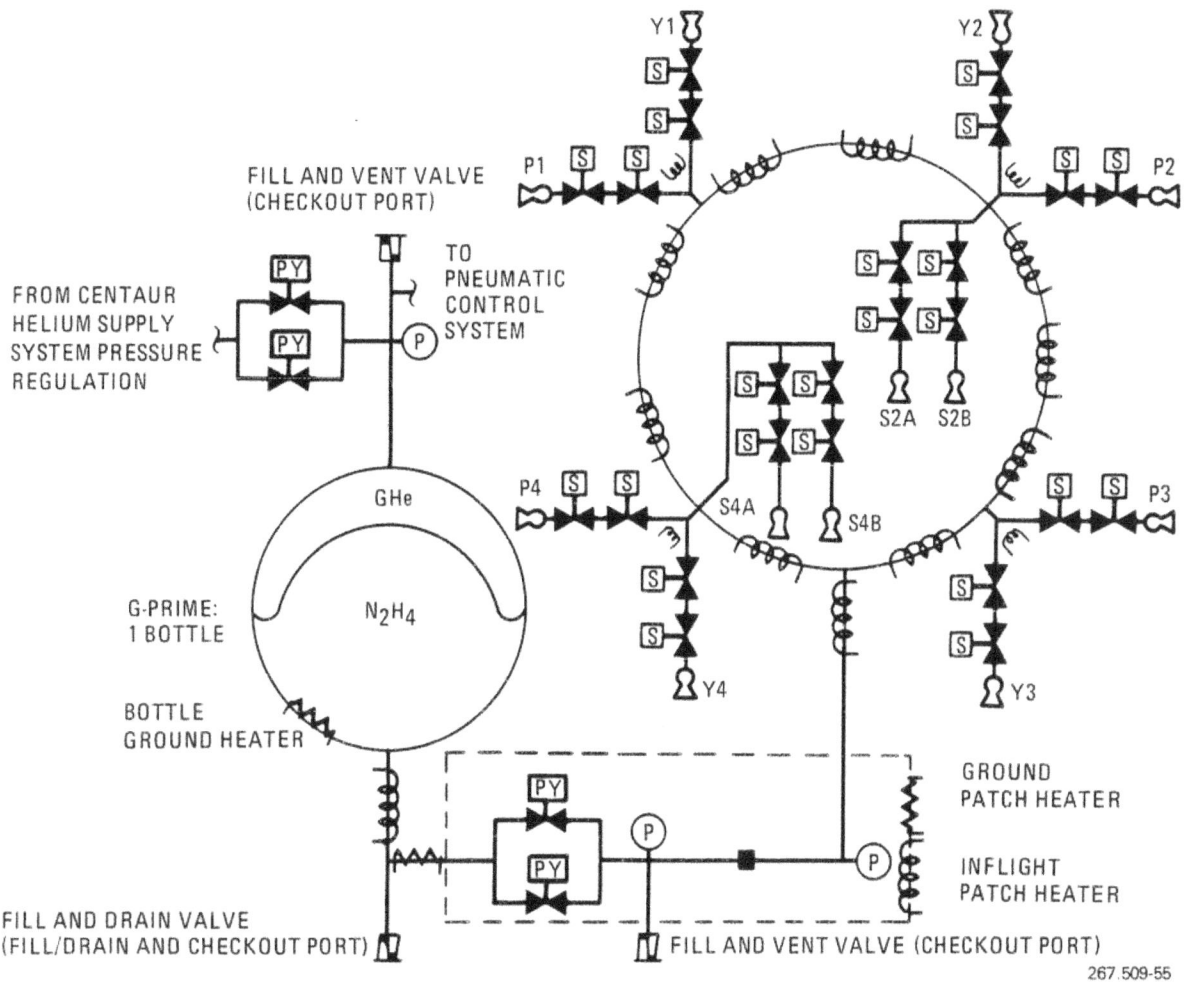

Figure 3-44. Reaction Control System

e. Centaur Vent System (Figure 3-45). Each propellant tank vent line contains a spring loaded open, pneumatically actuated closed, shutoff valve.
Each vent line runs from the shutoff valves aft to a disconnect in the
respective Centaur/CISS fuel and oxidizer disconnect panels. These vent
shutoff valves close the Centaur vent system prior to separation from the
CISS. The disconnects contain a self-sealing poppet to provide backup
shutoff capability after disconnect panel separation. The GH_2 vent
shutoff valve also serves to isolate the primary GH_2 vent system from
the on-orbit lower flow rate thermodynamic vent system (TVS).

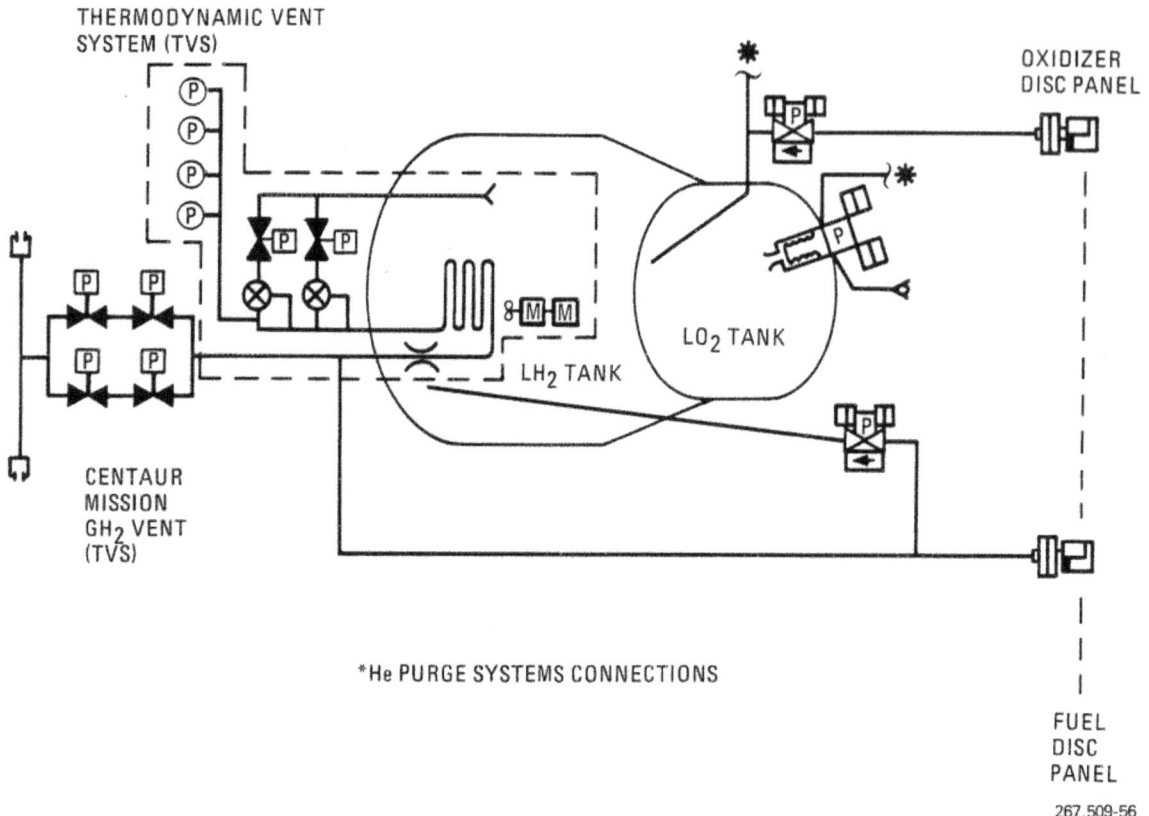

Figure 3-45. Centaur Vent Systems

The TVS is used for GH$_2$ vent control while in a zero-g environment.
This system provides for on-orbit predeployment venting through the
Orbiter fuel midbody dump port and freeflight venting through the Centaur
Mission Balanced Thrust Vent System. The system consists of a heat
exchanger with a parallel set of throttling regulators and integral
shutoff valves at the heat exchanger inlet and four series/parallel valves
downstream of the heat exchanger. All valves are pneumatically actuated.
The pneumatic supply to the downstream set of valves is isolated by pyro
valves to preclude venting into the cargo bay before Centaur deployment.
The TVS also contains a dual, electric motor driven pump to circulate
propellant through the heat exchanger and to maintain the LH$_2$ propellant
mixed to minimize the need for venting. This single-failure tolerant vent
system has been sized to maintain propellant tank pressure control while
in the closed door payload bay environment, which represents the maximum
boiloff condition.

The LO$_2$ tank contains a pneumatically operated JPM to accomplish the
mixing function. The LO$_2$ propellant thermal energy storage potential
(in the payload bay environment) can absorb all energy input if the
propellant is adequately mixed.

f. Centaur Fill/Dump System. The fill/dump system is designed to ensure
    Centaur G-prime compatibility with all Shuttle abort modes that occur
    before vehicle deployment. The system has been sized to provide
    propellant dump capability within 250 seconds - the minimum time allowed
    during a return to launch site abort. With this system, a simultaneous
    dump of LH_2 and LO_2 can be accomplished safely while the Orbiter is
    above 110,000 feet altitude, which corresponds to an ambient pressure less
    than 0.1 psia. Extensive testing has demonstrated that a hydrogen-oxygen
    mixture will not ignite at pressures below 0.1 psia.

    The Centaur fill/dump system is shown in Figure 3-46. This high flow
    capability, foam-insulated duct system and "normally closed" pneumatically
    actuated dump valves interconnect the LH_2 and LO_2 propellant tanks to
    a self-sealing disconnect in the respective Centaur/CISS fuel and oxidizer
    umbilical panels. The dump valves contain dual pneumatic actuators to
    assure opening capability. Propellant loading and draining are accom-
    plished through the same system during preflight operations. The self-
    sealing disconnects provide single-failure tolerance against inadvertent
    dumping after Centaur separation.

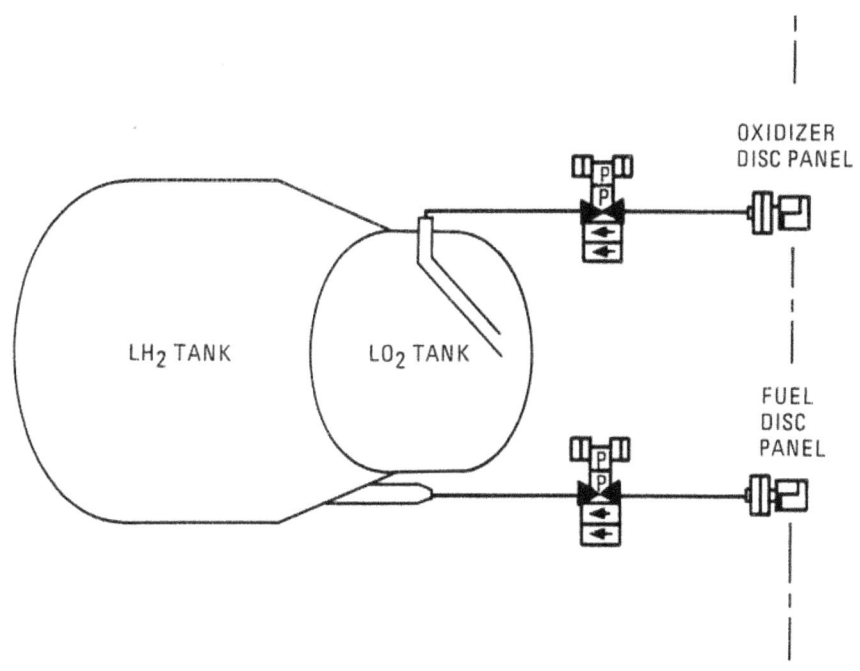

267.509-57

Figure 3-46. Fill/Dump Systems

g. Riseoff Fluid Disconnect System. There are two fluid system disconnect
    panel pairs - one in Quad I containing the hydrogen lines and one in Quad
    II containing the oxygen lines. The use of riseoff disconnects ensures
    fluid system continuity for venting of dumping until separation from the
    Orbiter. The asymmetrical separation impulse from the disconnects,
    although small, is considered in the deployment spring design. Figure
    3-47 shows the general arrangement of the panels with the disconnect sizes
    and fluid system functions indicated.

    The 5.5-inch fill/drain and dump disconnects and the 2.5-inch tank vent
    disconnects will be identical on both panels and will be qualified for
    both liquid hydrogen and oxygen service. A conceptual sketch of the
    5.5-inch disconnect is shown in Figure 3-48. The 2.5-inch disconnect is
    of similar design. The inner bellows provide the initial sealing force on
    the spherical seat and provides for misalignment and relative motion
    during ascent. The outer bellows provides a chamber for helium purge and
    leakage containment. The remaining disconnects are not cryogenic.

    To deploy the Centaur, the Super*Zip is fired and deployment springs
    accelerate the vehicle away from the deployment adapter. This motion
    separates the disconnect halves mounted on the Centaur panel from the
    mating halves mounted on the deployment adapter panel allowing the valve
    poppets to close.

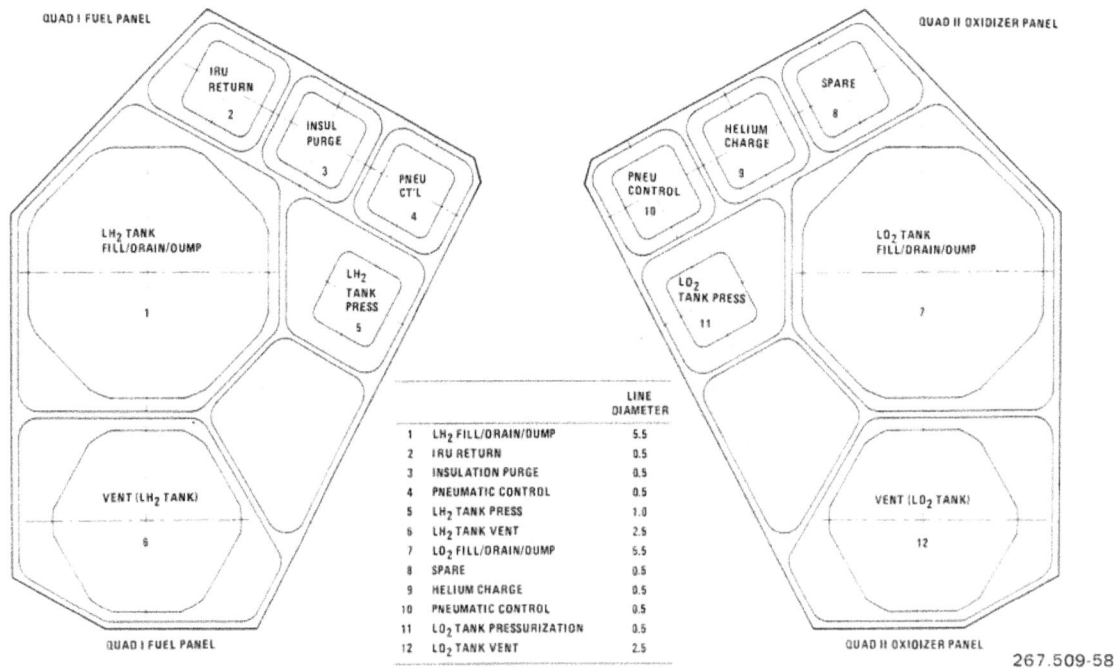

Figure 3-47. Centaur/CISS Fuel and Oxidizer Disconnect Panels

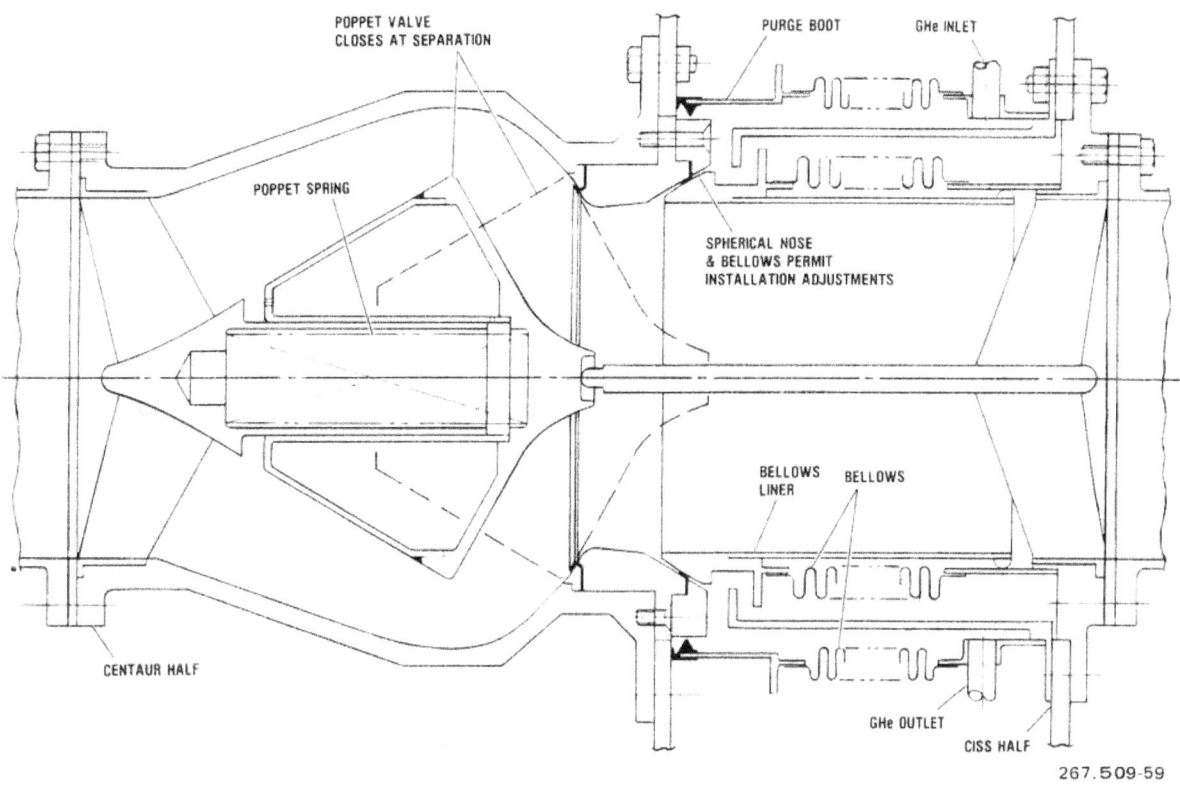

POPPET VALVE
CLOSES AT SEPARATION

PURGE BOOT   GHe INLET

POPPET SPRING

SPHERICAL NOSE
& BELLOWS PERMIT
INSTALLATION ADJUSTMENTS

BELLOWS
LINER   BELLOWS

CENTAUR HALF

GHe OUTLET

CISS HALF

267.509-59

Figure 3-48.   5.5-Inch Disconnect

3.3.4.2 <u>CISS Fluid Systems</u>. Major elements of the CISS fluid systems
required to interconnect the Centaur vehicle and the Orbiter interfaces, and
the systems required to provide the necessary safety requirements are
described in the following sections.

a.   <u>Propellant Tank Vent System (Figure 3-49)</u>. The CISS vent systems consist
     of disconnects, valves, and ducts connecting Centaur/CISS fuel and
     oxidizer disconnect panels and the associated Orbiter overboard ports.
     Both the GH_2 and GO_2 vent systems use similar ducting containing three
     gimbals to connect the deployment adapter half of the disconnects in the
     fuel and oxidizer panels to the hard-mounted ducting on the CISS (Figure
     3-12). These gimbal duct sections allow the Centaur to be rotated 45
     degrees to the deployment position with the vent systems still connected
     and fully operational, maintaining the required vehicle safety controls.

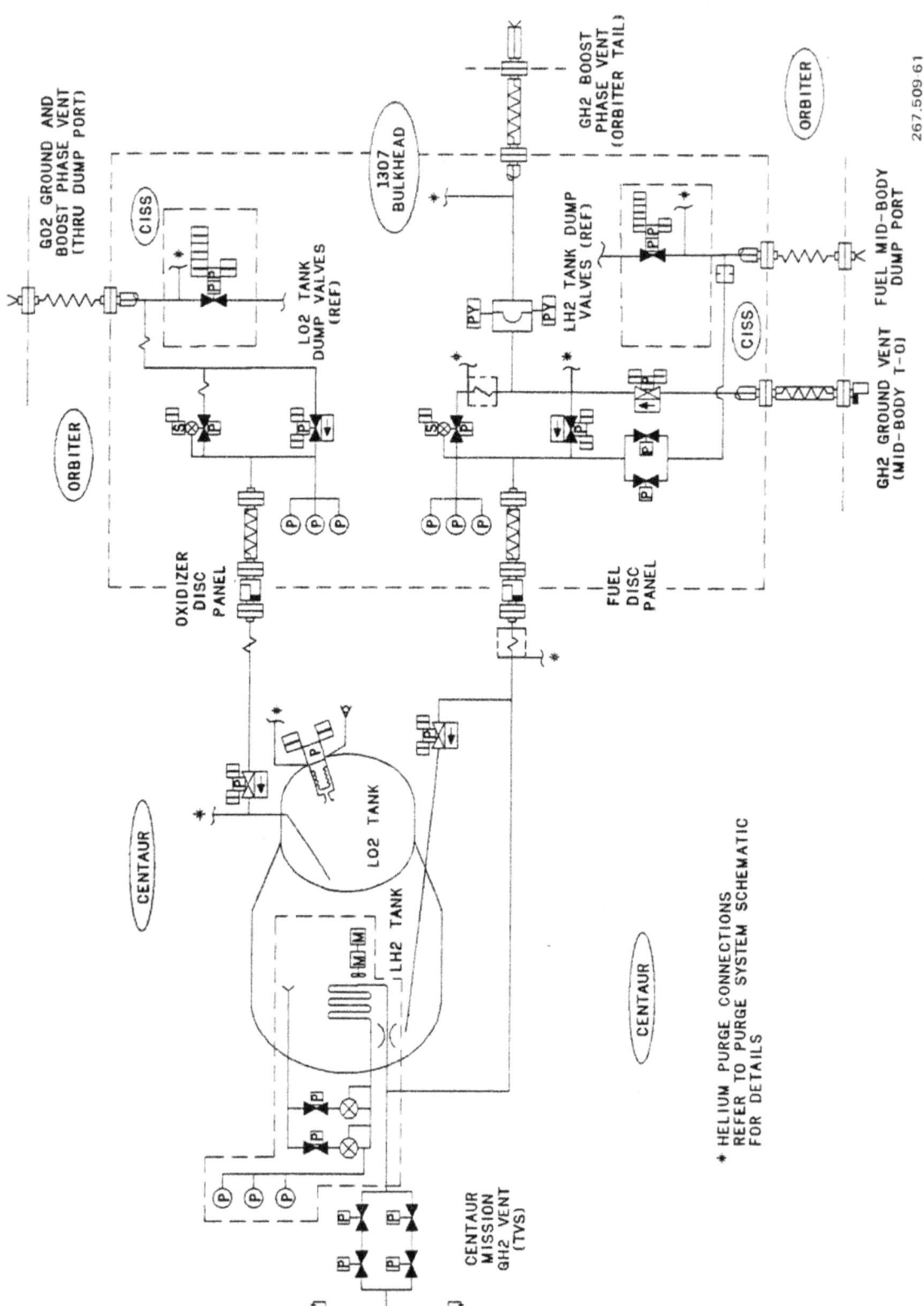

Figure 3-49. CISS Vent System

267.509 61

The GH_2 system branches on the CISS to three Orbiter overboard paths: the port side GH_2 ground vent (midbody T-0), fuel midbody dump port, and GH_2 boost phase vent (Orbiter fail). The ground and ascent vent path branches through two parallel Centaur vent valves. One valve is a primary mechanical self-regulating vent valve with solenoid lockup capability used for controlling ground and ascent venting. The valve is the same as that used on all previous Centaur vehicles. The secondary valve, in parallel with the primary, is a pneumatically actuated open and spring-loaded closed shutoff valve which provides backup ground and ascent vent capability. All GH_2 ground venting is ducted from the CISS to the Orbiter port side midbody T-0 disconnect panel. A spring-loaded normally open and pneumatically actuated closed valve, located on the CISS between the primary/secondary vent valve outlet leg and T-0 disconnect, provides backup shutoff capability for LH_2 tank ground venting after liftoff.

In-orbit GH_2 venting from the Centaur TVS is controlled by two parallel valves mounted on the CISS. These valves are normally spring-loaded closed and require pneumatic pressure to open. Pyro isolation valves (not shown) inhibit pneumatic actuation of these valves while on the launch pad. Venting while on orbit is through the Orbiter fuel midbody dump port.

The third GH_2 vent path directs GH_2 venting from either the primary or the backup ground ascent valves to the boost phase vent located near the Orbiter tail. Venting through this path is inhibited until after launch by a pyro-operated frangible valve to prevent overboard venting while on the launch pad.

On the ground and during ascent, the GO_2 system vents through the Orbiter starboard side dump port. The ground and ascent vent control of the LO_2 tank is similar to the GH_2 system. A primary self-regulating valve maintains tank pressure with a secondary backup valve in parallel with the primary valve. There is no requirement for on-orbit venting of the LO_2 tank.

b. Fill/Drain and Dump System (Figure 3-50). The CISS LH_2 and LO_2 fill/drain and dump systems are schematically identical systems. However, the LO_2 system has a 3.5-inch valve in the fill/drain line to preclude surge pressures from entering the Orbiter line. The dump systems are sized to provide the requirement for dumping propellants in 250 seconds during a RTLS abort operation. Both the LH_2 and LO_2 systems use similar ducting continuing through gimbals that connect the deployment adapter half of the disconnect to the hard-mounted ducting on the CISS (Figure 3-29). Again, as in the vent systems, these gimbal duct sections allow the Centaur to be rotated 45 degrees to the deployment position with the dump system still connected and fully operational, maintaining the required vehicle safety controls.

267.509 62

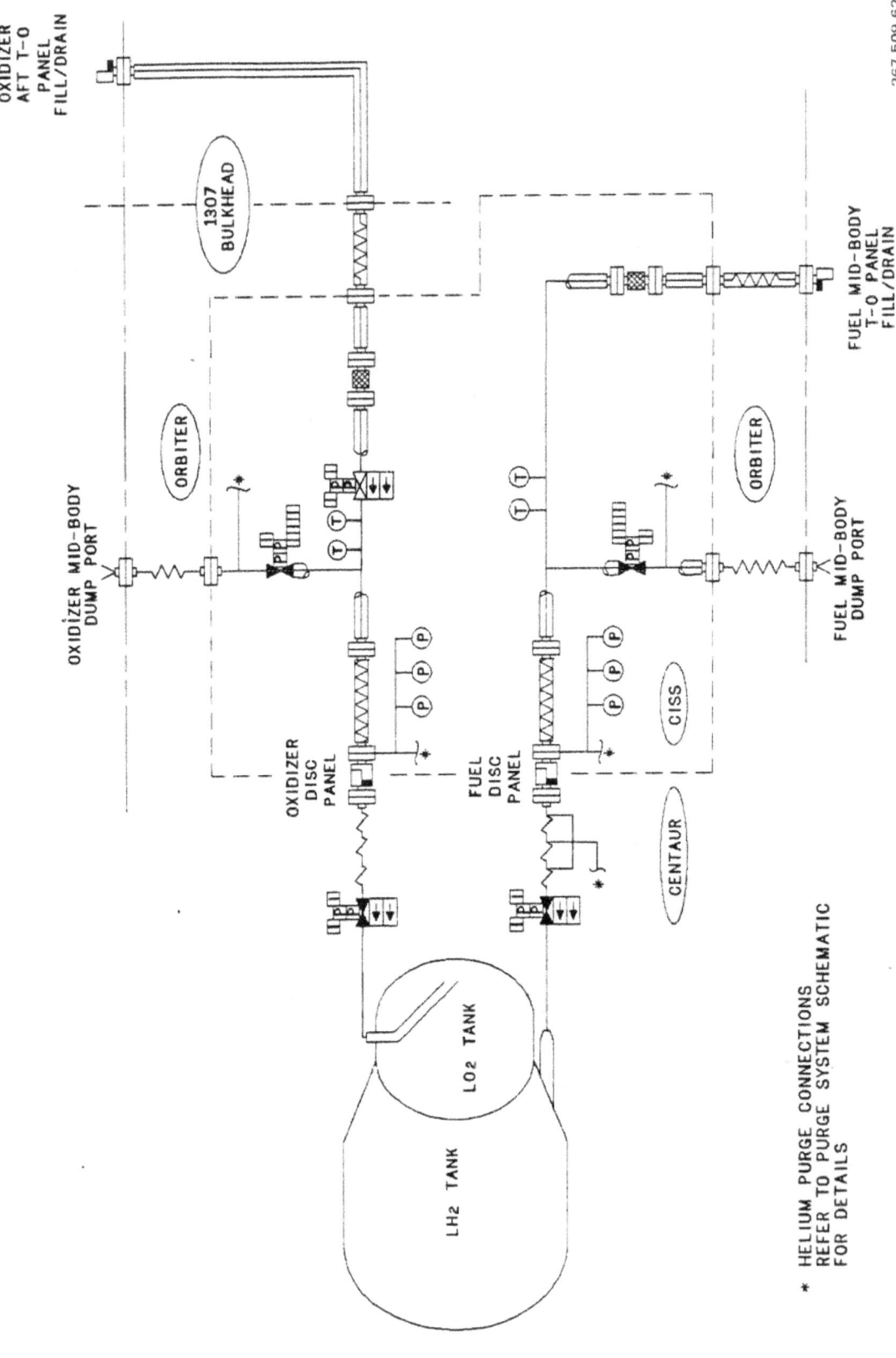

Figure 3-50. Centaur/CISS Fill/Drain/Dump System

The LH$_2$ duct system is routed from the three gimbal duct on the CISS through a manifold to Orbiter ducting connected to the fuel midbody T-0 panel and dump port. The CISS dump valve is a 5.5-inch dual actuator pneumatically open, spring-loaded closed valve with pyro valve pneumatic isolation (not shown) to prevent inadvertent opening of the valve during LH$_2$ tanking on the launch pad.

The LH$_2$ fill and drain line is routed from a CISS manifold which is interconnected with the three gimbal duct, to Orbiter lines connected to the Orbiter fuel midbody T-0 panel. The CISS fill and drain line includes a filter to protect CISS and Centaur ducting and components from contamination. Similarly, a LO$_2$ 3.5-inch valve protects the Orbiter line from pressure surges. Flow control during fill and drain operations is provided by the GSE.

The LO$_2$ dump system is identical to the LH$_2$ dump system except the LO$_2$ exits through the Orbiter oxidizer midbody dump port.

The LO$_2$ fill and drain line is also similar to the LH$_2$ system, except the CISS ducting connects to the Orbiter oxidizer aft T-0 fill/drain port.

The dump systems also provide additional paths for tank pressure relief during a contingency operation.

Purges are provided at the LH$_2$ and LO$_2$ dump ports to prevent "cryopumping" into the open dump ports while on the launch pad.

c. Pneumatic Systems. CISS pneumatic systems consist of the propellant tank pressurization system, pneumatic actuated valve control systems, helium purge systems, and helium supply system.

These interconnected systems are described individually in the following section.

(1) Pressurization System (Figure 3-51). The CISS pressurization system consists of a valve module containing two parallel legs of two solenoid valves in series for each propellant tank. The two legs contain identical flow control orifices so any one leg will provide adequate pressurization for liftoff. Two legs will provide adequate backup to the Centaur pressurization valves for abort dump pressurization.

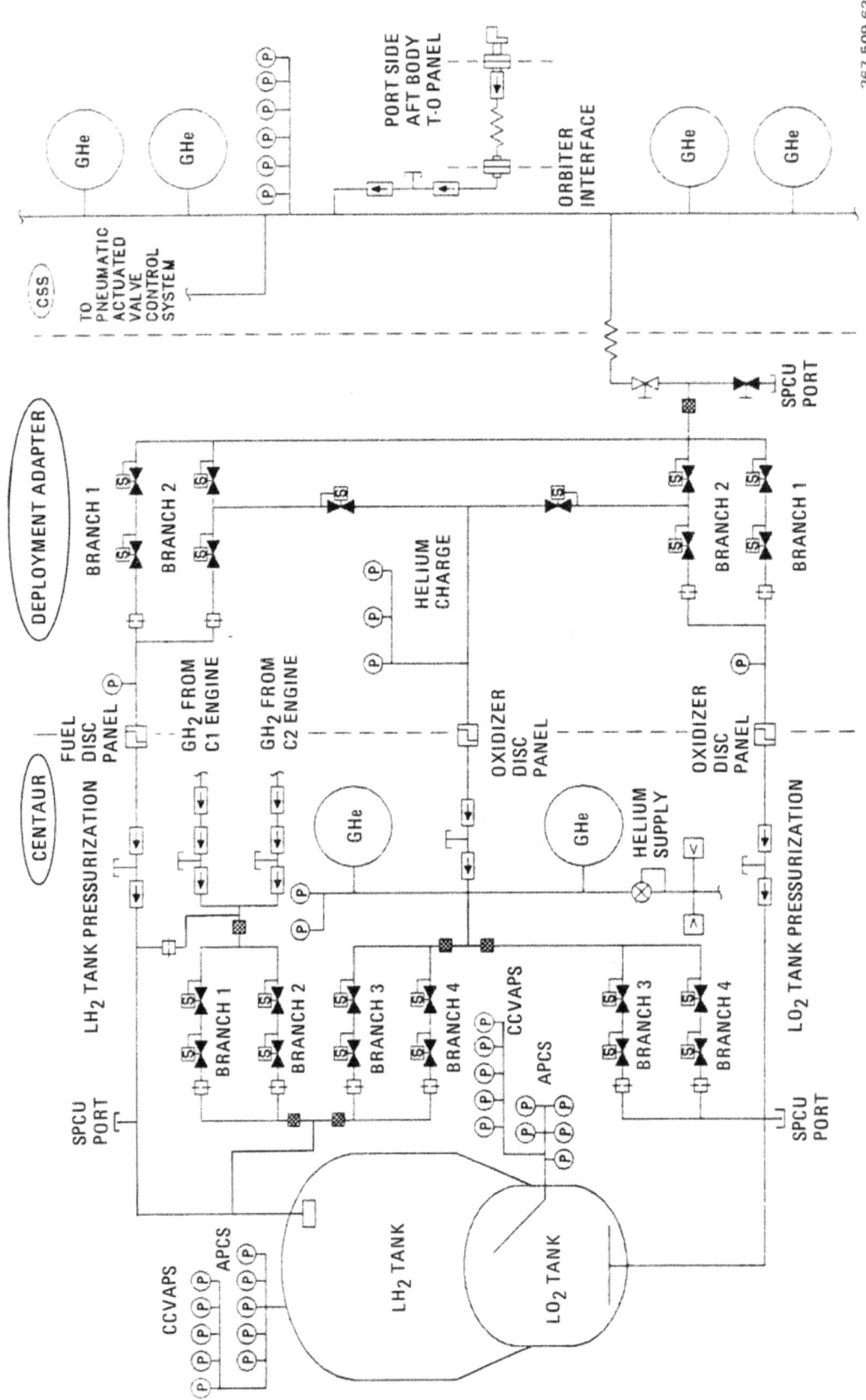

Figure 3-51. Propellant Tank Pressurization System

The system is single-failure tolerant against inadvertent pressurization and in providing low-flow and in providing high-flow when combined with the Centaur system. The valve modules are mounted on the
interior surface of the deployment adapter with a supply line
connecting the respective valves to disconnects in the Centaur/CISS
fuel and oxidizer umbilical panels. All system permanent tube joints
are welded and all separable connections use welded "Dynatube"
fittings. A quick-disconnect port with a manual shutoff valve and
dual sealed cap permits connecting the standby pneumatic control unit
to the helium supply line whenever the helium control skid is not in
operation and to provide helium supply for system checkout.

(2) Pneumatic Activation Valve Control System (Figure 3-39). Three
parallel regulators assure a two-failure tolerant nominal 450 psig
activation supply to any one of three separate actuation supply
branches. Two parallel branches each provide actuation pressure to
the dump valves and zero-g vent shutoff valves. One branch provides
actuation pressure to the CISS secondary LO_2 and LH_2 vent valves
and the LH_2 ground backup vent valve. The third branch supplies the
various purge systems and routes to the Centaur vehicle through two
separate parallel paths. Actuation pressure to the dump valves and
on-orbit zero-g vent valves are pyro isolated to prevent inadvertent
actuation while on the launch pad. Each branch can be cross-connected
to the other branches in the event of failure or for an abort return.

(3) Purge System (Figure 3-40). The CISS purge system is supplied from
the two-failure tolerant pressure regulation system described above.

A network of six solenoid valves provides two parallel systems, each
of which can provide two different purge rates to the Centaur LH_2
tank insulation blanket through a disconnect in the Centaur/CISS fuel
umbilical panel. An additional network of four solenoid valves
provides a purge to the Centaur engines and LO_2/LH_2 umbilical
disconnects. Another six solenoid valves control purges to the LO_2
and LH_2 fill/drain system and LH_2 ground vent. Vent lines route
the helium purge and any LO_2 and LH_2 system leakage to respective
overboard vent ports in the Orbiter midbody fuel T-0 panel, LH_2
inflight dump, and the starboard GO_2 ground-boost vent (midbody
dump).

The line supplying the LH_2 tank insulation purge contains a capped
port to permit connecting the standby pneumatic control unit to supply
this purge when the automatic avionics control system is not in
operation.

(4) Helium Supply System (Figure 3-41). A network of twenty 4,650 cubic
inch (22-inch diameter) Kevlar-overwrapped helium storage spheres is
mounted on the CISS structure to provide all G-prime helium requirements while in the Orbiter. The capability of this system is sized to
permit dumping of propellants in 250 seconds during a RTLS abort. The

two CISS helium charge line check valves and the Orbiter supplied
helium charge check valve provide three in-series shutoff devices at
liftoff. Charging of the Centaur and CISS helium storage bottles is
controlled by the GSE.

3.3.5 AVIONICS SYSTEM. The Centaur G-prime avionics system is essentially
the Centaur D-1A system with modifications to satisfy communications and
safety requirements while accomplishing the baseline, one-burn mission.
Figure 3-52 shows major functional elements and signal interconnections.

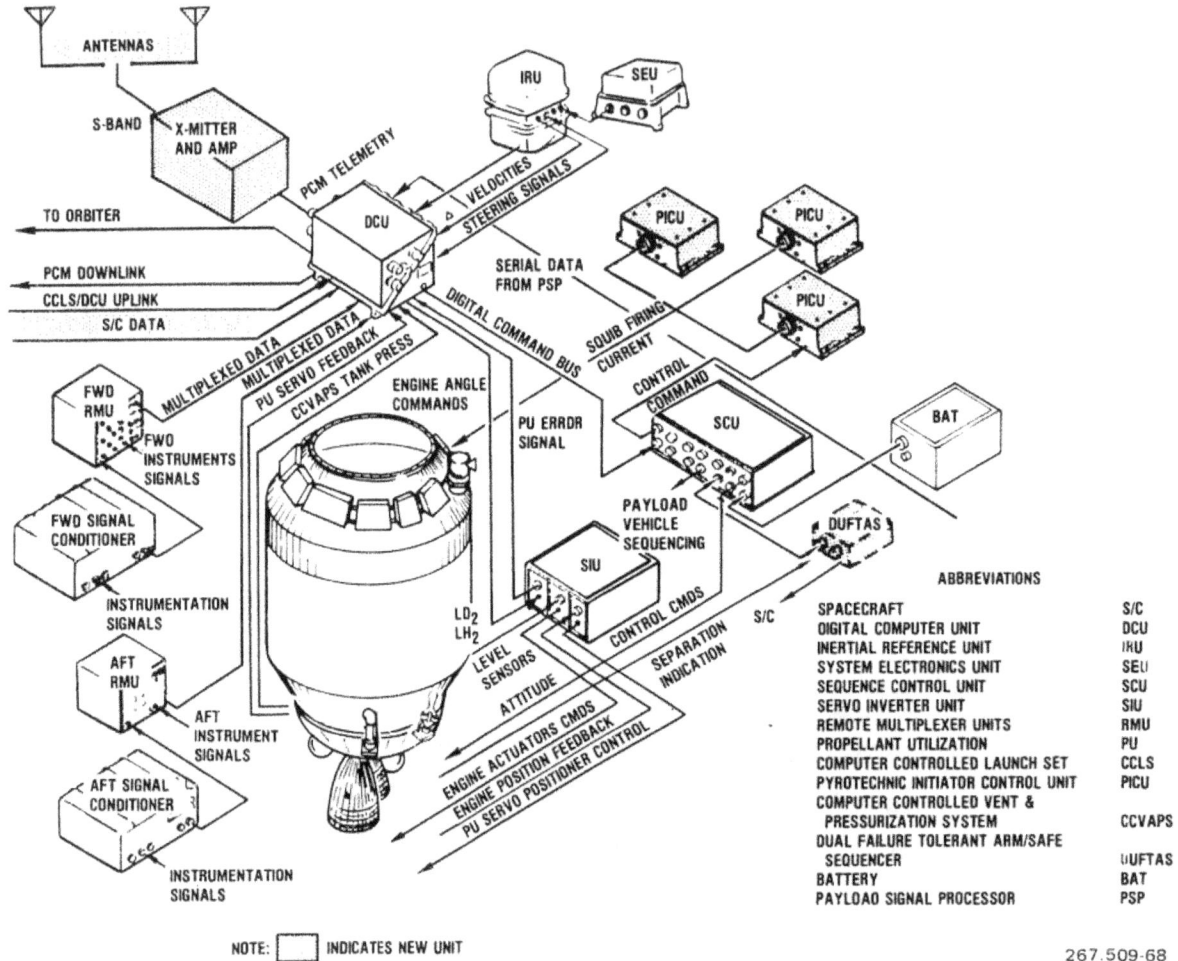

Figure 3-52. Centaur G-Prime Avionics

3.3.5.1 Centaur Vehicle Avionics

a. Digital Computer Unit. The Teledyne Systems Co. digital computer unit
    (DCU) is a 16,384-word, 24-bit, random access core memory computer. Its
    memory is divided into 12,288 words of nonalterable memory and 4,096 words
    of alterable memory. The DCU receives incremental time and velocity
    pulses from the inertial measurement group, converts analog dc to digital
    form, receives formatted data from remote multiplexers and external
    sources, responds to discrete inputs, and acts on priority interrupts.

The DCU output converts digital values to both ac and dc signals, formats PCM data, and provides parallel discrete commands.

The input/output capability provides hardware compatibility with Orbiter differential inputs, communication with the Orbiter or the ground, in addition to input/output interfaces with other Centaur GN&C functions.

The Centaur G-prime DCU incorporates minor modifications to the Centaur D-1A unit to accommodate a different telemetry system.

b. Guidance and Navigation. The guidance and navigation functions are implemented using a Honeywell inertial measurement group (IMG) for measurement of vehicle acelerations (ΔV pulses), and the DCU for computation of vehicle position and velocity and generation of the required steering signals. The IMG is composed of an inertial reference unit (IRU) and its associated power supplies housed in the system electronics unit (SEU). The IRU contains a four-gimbal, gyro-stabilized platform that supports three orthogonal, pulse-rebalanced accelerometers.

A gyrocompassing mode is used for initial azimuth alignment of the inertial element for Shuttle missions.

Although IRU accuracy is sufficient to meet the Galileo mission requirements by navigating from the ground up without external navigation or attitude assist, extended periods in the Orbiter payload bay may necessitate an attitude update. Navigation update capability from the Orbiter PSP is available when required for specific missions.

The Centaur G-prime IMG is a modified Centaur D-1A IMG to improve gyro torquing command accuracy to accomplish attitude update and azimuth alignment.

c. Control. The control functions -- Centaur main engine thrust vector control and coast-phase attitude control -- are performed on the basis of analog vehicle attitude errors received by the DCU from the IMG. The main engine thrust vector control signals are generated by the digital autopilot software in the DCU, which accepts the IMG attitude errors, differentiates them to obtain vehicle rates, and computes the desired engine actuator commands. The analog engine actuator commands are sent from the DCU to the servo inverter unit (SIU). Here they are power-amplified and the engine position control loop is closed by position feedback signal from an actuator transducer. The attitude control signals during coast phase are also generated by DCU computations, which use the IMG attitude errors as inputs. The coast-phase-attitude engine commands are digital on-off commands sent from the DCU to the sequence control unit (SCU). SCU relays provide the attitude control engine switching.

The Centaur G-prime control system is essentially the same as Centaur D-1A avionics control systems. It is safety-related except for the DUFTAS.

The DUFTAS provides dual-failure tolerance against inadvertent initiation of all safety-related functions until deployment of the Centaur is complete and the separation distance is sufficient. The DUFTAS contains three timers initiated by breakwires in the separation disconnects. Figure 3-53 indicates the functional mechanism for providing two-failure tolerance.

In addition to flight controls and safety functions, Centaur avionics provide command outputs based on a serial uplink from the Orbiter (Figure 3-54). This command link is available only during the attached mode and initiates discrete commands through the Centaur DCU/SCU to the spacecraft.

d. Sequencing. The SCU contains the logic to decode a 22-bit parallel DCU sequencing command and 96 magnetic latching relays. The SCU provides all sequencing commands to vehicle systems and 16 discrete commands to the spacecraft.

e. Propellant Utilization Management. This system increases vehicle performance by propellant residual management and by controlling the engine mixture ratio.

For Centaur G-prime, the Centaur D-1A capacitance probe used to sense liquid levels will be changed to conform to the modified tank shape. Part value constants in the bridge detector circuits will be changed to maintain the present 5:1 mixture ratio control.

f. Computer Controlled Vent and Pressurization System (CCVAPS). CCVAPS controls pressurization of the Centaur hydrogen and oxygen tanks and venting of the hydrogen tank, and selects the transducer to be used for tank pressure control.

During the coast phase before main engine start, hydrogen tank pressure is controlled by CCVAPS using the TVS. During this period, CCVAPS also controls the oxygen tank pressure using the oxygen tank JPM.

Pressurization of the propellant tanks begins shortly before main engine burn and continues throughout the burn. There are two pressurization branches for each tank: one active and one backup. CCVAPS tests the active pressurization branch for failure and, in the event of failure, switches to the backup branch.

Throughout the mission, CCVAPS tests each tank's pressure transducers for tank pressure control. Hydrogen tank structural safety control (using the hydrogen vent shutoff valve) and intermediate bulkhead safety control (using the active pressurization or venting system) are also active throughout the mission.

267.509.69

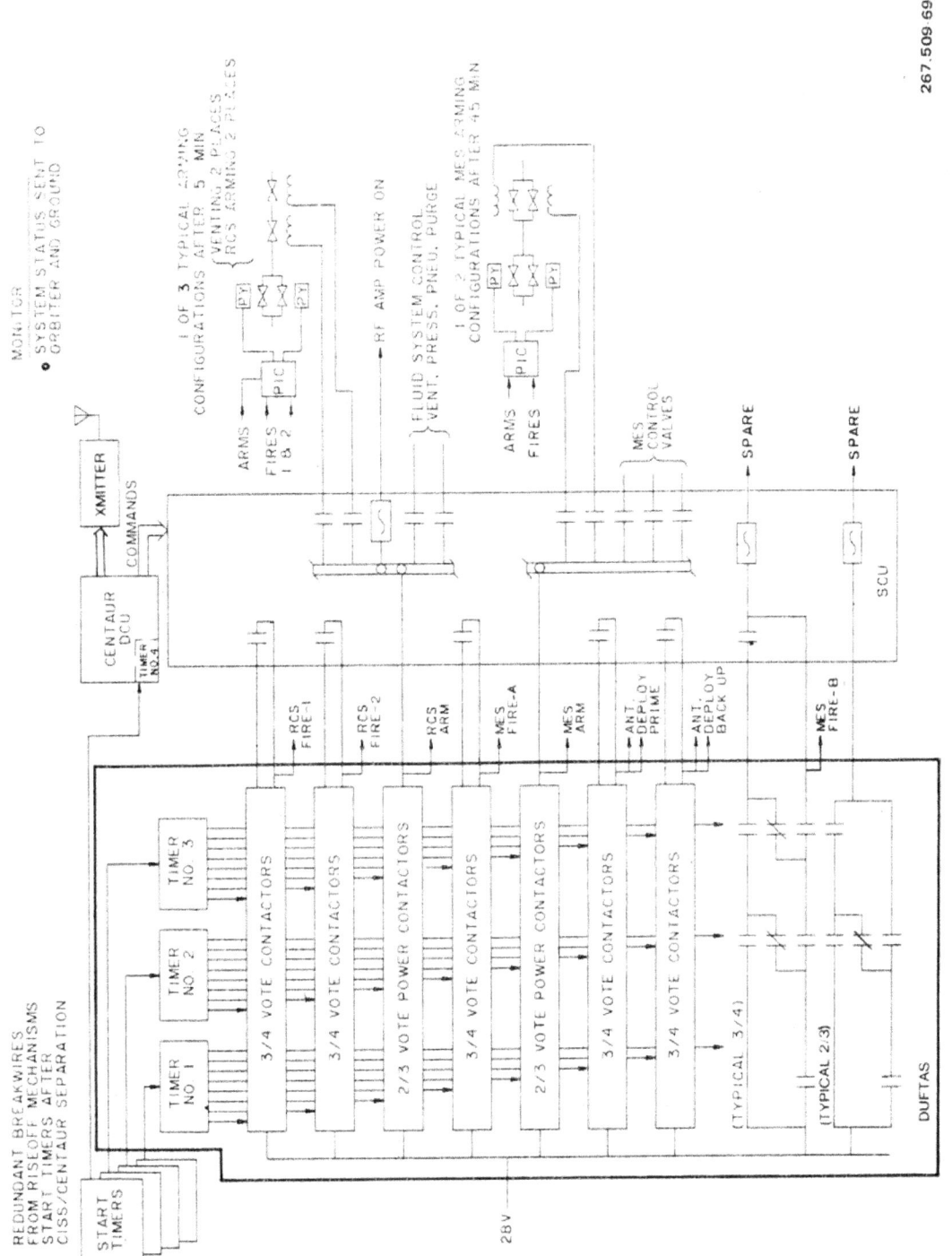

Figure 3-53.   Failure Tolerant Timed Arming Sequence

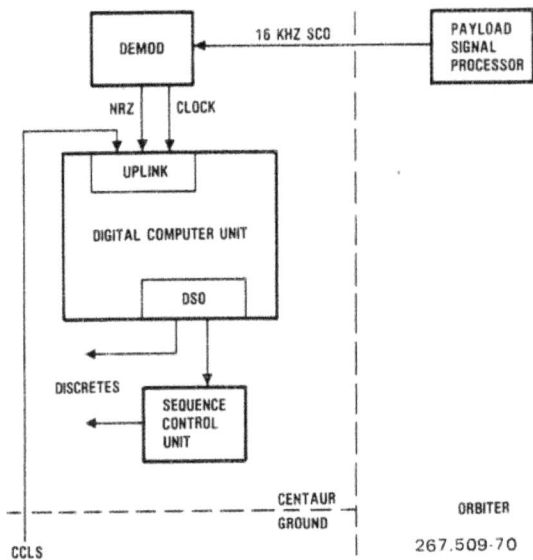

Figure 3-54. Baseline Centaur G-Prime Command Link Paths

g. <u>Instrumentation</u>. The Centaur G-prime instrumentation system (Figure 3-55) collects data, and conditions and converts it to a PCM bit stream. The system provides this information before launch and throughout all phases of Shuttle and Centaur flight phases. The instrumentation system consists of vehicle transducers, harnesses, signal conditioners, remote multiplex units (RMU), and the PCM central control unit (CCU), supplied with and located in the DCU.

The system is portioned into two remotely located groups of equipment: (1) a set of transducers, a signal conditioner, and a RMU on the forward end of the Centaur, and (2) an identical set at the aft end of the Centaur.

The signal conditioner produces transducer excitation voltages and normalizes the measurement signals to acceptable ranges for the RMU.

RMU addressing is under format control of the CCU. The CCU also integrates PCM data from the spacecraft payload with internal Centaur vehicle data.

The composite PCM data is sent over the following links:

- Prelaunch data to the GCS.

- Attached data to payload data interleaver (PDI) and payload recorder.

- Detached data to Orbiter or ground via TDRSS-compatible transmitter.

Safety function transducers (insulation blanket differential pressure and tank pressure) are directly excited by the 28 Vdc bus and hardwired directly to the CISS safety control system. Each safety function transducer output is independently signal conditioned.

All elements of the system are similar to the Centaur D-1A instrumentation system.

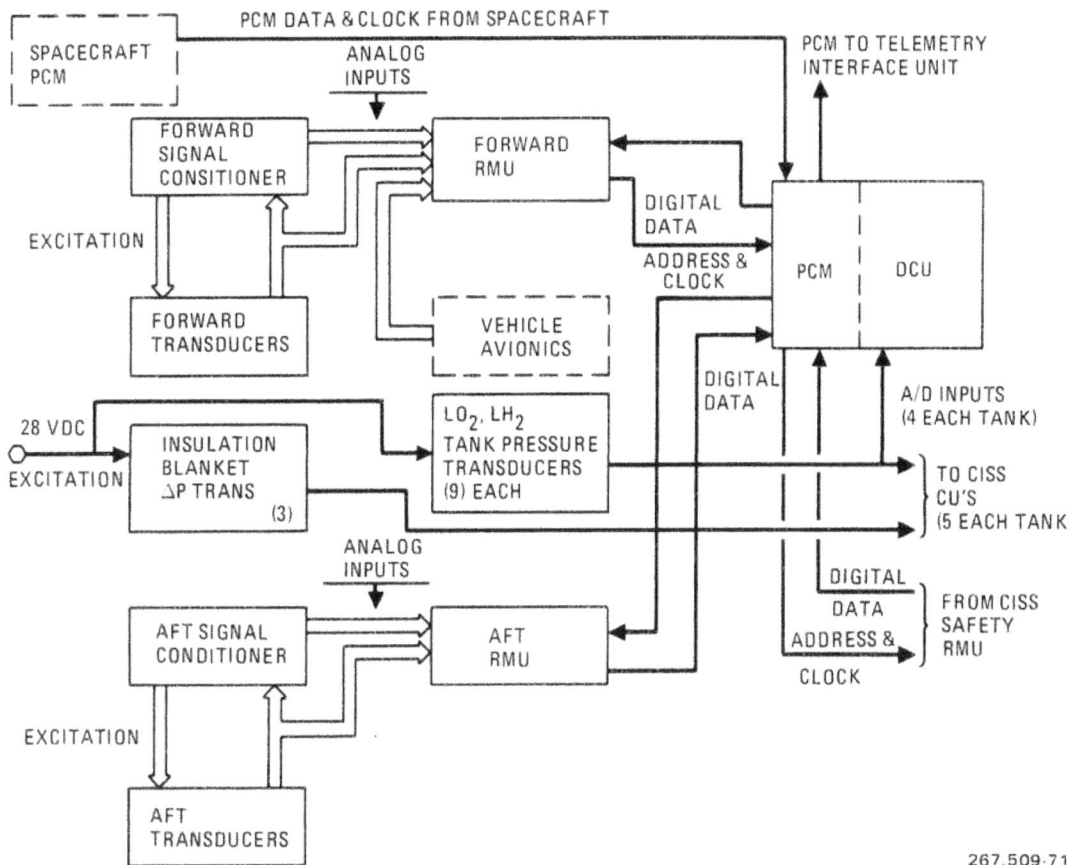

Figure 3-55. Centaur G-Prime Instrumentation System

h. <u>Telemetry</u>. The baseline Centaur G-prime telemetry system (Figure 3-56) is used to communicate to the ground and Orbiter from the Centaur, both in the attached and deployed modes. Downlink transmissions are required for 16 kbps of PCM data to the Orbiter and 16 or 32 kbps to the ground.

Major telemetry hardware elements are the S-band transmitter, RF power amplifier, RF coaxial switch for antenna selection, two antennas for full spherical coverage, waveguide assemblies, and coaxial cable assemblies.

The telemetry system is compatible with the Orbiter Payload Interrogator and the NASA TDRSS.

The transmitter responds to commands from the Centaur SCU to reconfigure, provides self-monitoring operation modes, and provides self-monitoring status signals to the instrumentation system. The transmitter is switched to the appropriate S-band antenna, under DCU control. DCU decisions are based on the known relative position of the intended receiver and the attitude of the Centaur vehicle. The transmitter and power amplifier are activated when the SCU closes switch 96 under DCU control. The DCU also controls the RF coaxial switch that selects the proper antenna for S-band signal transmission. The telemetry system interfaces with the GN&C system to assure proper antenna selection.

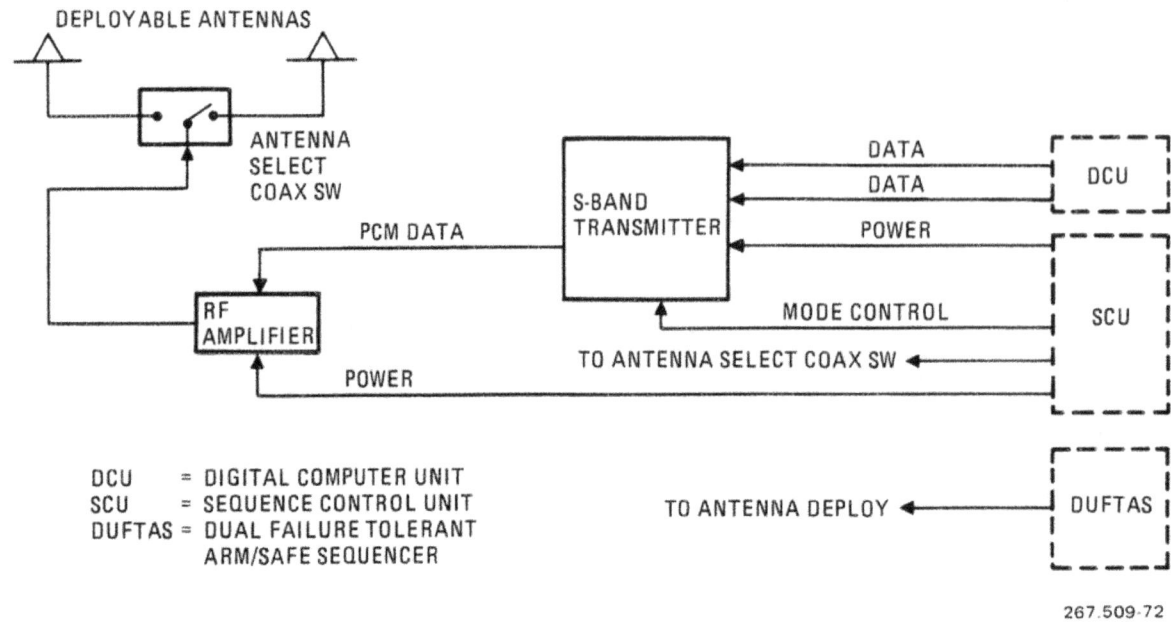

Figure 3-56. Centaur G-Prime Telemetry System

i. **Electrical Power.** The Centaur G-prime electrical power and distribution
system consists of a silver-zinc primary battery, a power changeover
switch to select between Centaur internal (battery) and external power,
riseoff umbilicals located on Centaur/CISS, military specification wire
and connectors comprising the distribution system. This is a derivative
of the flight-proven system flown successfully on Centaur D-1A. The
principal changes accommodate the larger equipment mount ring, redesigned
tank, and new equipment. Figure 3-57 is a functional diagram of the
Centaur G-prime electrical power system.

3.3.5.2 **CISS Avionics.** Five control processors with majority voting ensure
compliance with Orbiter safety requirements. Control of the systems at the
functional level with voting at the output plane simplifies the redundancy
management (Figure 3-58).

The CISS provides all electrical interfaces between Centaur and the Orbiter.
CISS also provides:

● Computer control for operational sequencing of all systems requiring
  multiple-failure tolerance.

● Electrical power and power control.

● Instrumentation and telemetry.

● Pyrotechnic control for Centaur separation.

● Centaur propellant-level tanking indications.

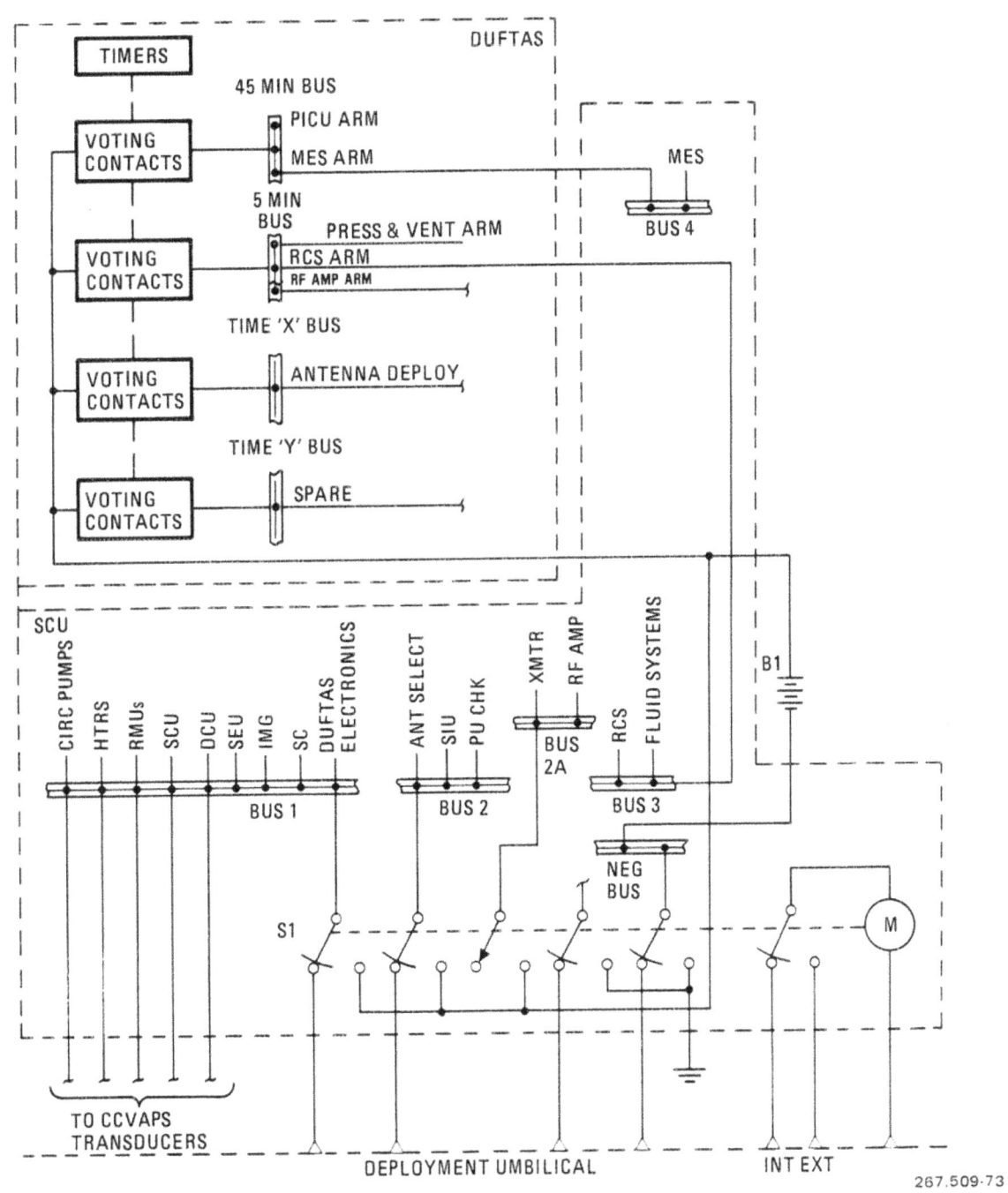

Figure 3-57. Centaur G-Prime Electrical Power System

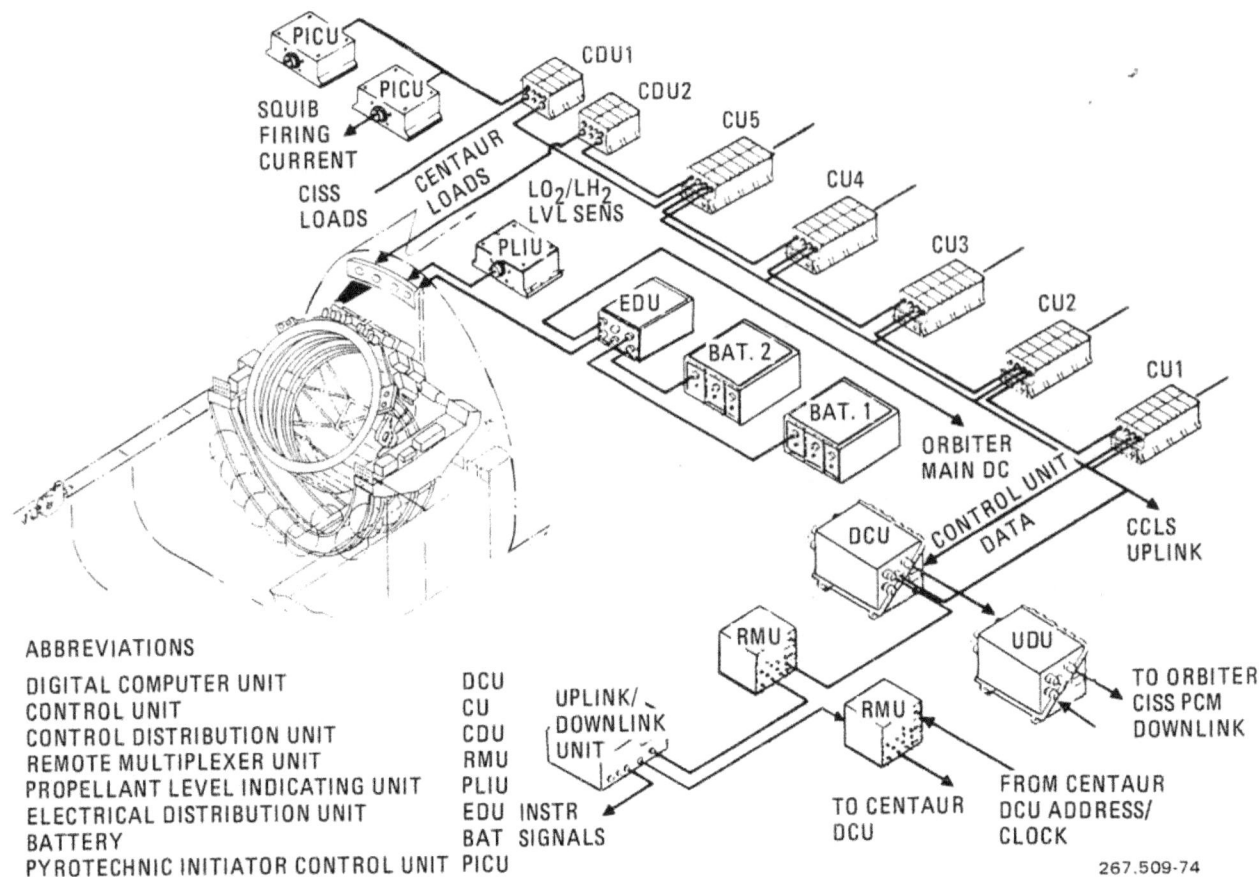

Figure 3-58. CISS Avionics

All harnesses providing Orbiter services to and from the Centaur cargo element (CCE), including the spacecraft, interface through the CISS. Standard Orbiter-provided interfaces are used for all uplink command and downlink data services, for crew monitor and control, and for hardwires to ground support equipment through the T-0 umbilical.

a. Control System. The CISS control subsystem governs the safety functions on CISS and Centaur (during predeployment operations) at a two-failure tolerant level for catastrophic functions. All systems are controlled at the functional level to eliminate extensive feedback networks required to detect failures at the component level.

The control system executes a majority vote at the output of each of the five independent control units. The vote is achieved at the output plane by interconnection of relay contacts, as shown in Figure 3-59. Voting at the power-load interface eliminates the necessity for failure detection within the electronic strings. This approach permits straightforward functional control laws for subsystem control and fault detection.

A Centaur DCU located on the CISS provides PCM data from all control units to the ground and to the Orbiter. The DCU is also used in the instrumentation system.

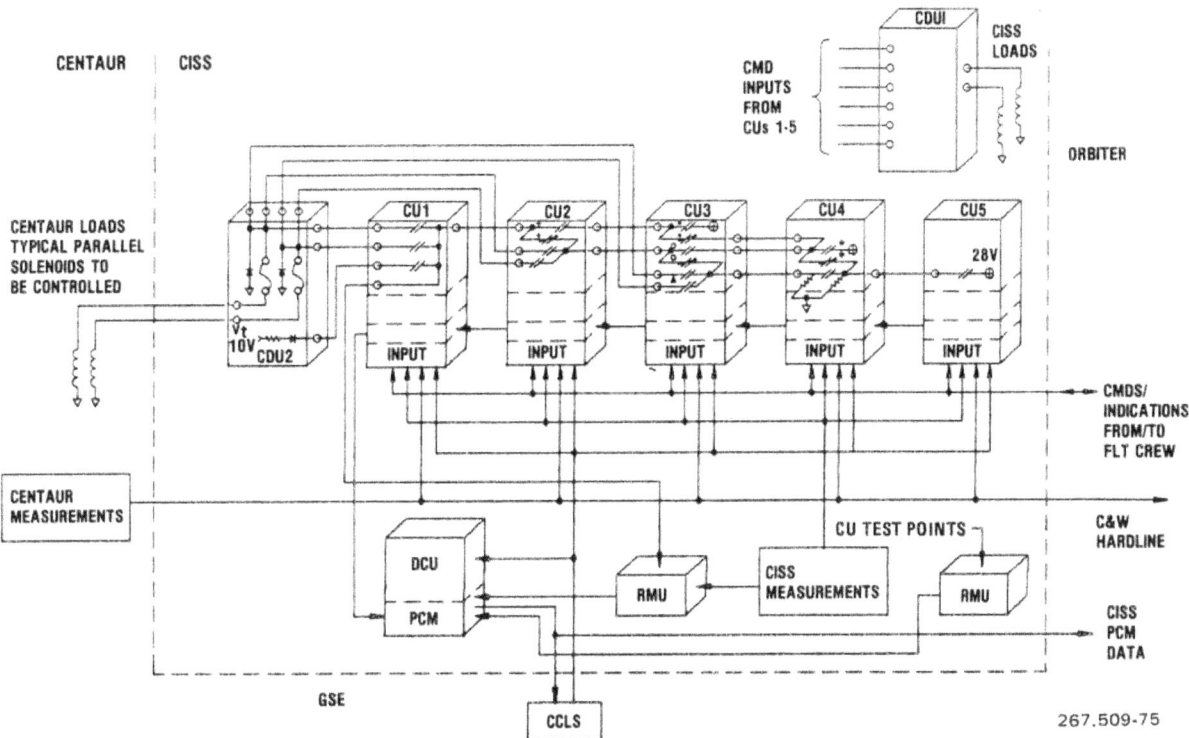

Figure 3-59.   Five-String Voter Ensures Two-Failure Tolerance

The relay networks are designed to provide individual load control from the CCLS.   Test points and instrumentation permit verification of proper functioning of individual control units (CU) and relay networks during all operations.

Command and data communication links provide individual control and monitoring of the CUs.   Commands to the CUs during ground operations originate at CCLS and are used to load flight-dependent constants, initiate ground operations, and execute prelaunch checkout programs. During flight, discrete commands originating at the switch panels will initiate in-line automated sequences such as the on-orbit deployment sequences.   The abort dump of propellants is planned to be initiated by discrete commands from the Orbiter's automatic abort sequence.

Downlink PCM data paths will be provided to the GCS and to the Orbiter for visibility into the state of the CISS.

Two data paths, one each from the CISS and Centaur, provide monitor capability for closed-loop ground checkout operations with the CCLS.   A third abbreviated data set for manual detanking is made available through the Orbiter.   The status of all systems will be available to the Orbiter and to the ground after liftoff via the three data paths.   The data will be used to verify proper system operation.   Some of the data are transmitted to the Orbiter's caution and warning system and serve as backup to the discrete caution and warnings generated by the CUs.

b. <u>Electrical Power System</u>. The CISS electrical power and distribution system consists of two backup silver-zinc primary batteries, redundant with
primary Orbiter power to provide two-failure tolerant power to CISS/
Centaur in the attached mode. Power control, diode isolation, and bus
excitation for both Centaur (external) and CISS loads are provided by the
electrical distribution unit (EDU). The Centaur umbilicals are zero-force
(nominal) riseoff disconnects. Military specification wire and connectors
are used in the harnesses. All of these assemblies/components are space
flight-proven or direct derivatives thereof.

c. <u>Instrumentation System</u>. The Centaur G-prime CISS instrumentation system
(Figure 3-60) follows the operating hardware, installation, and general
practice philosophy currently used on Centaur D-1A. A signal conditioner
and a RMU support the CISS measurements. A second RMU provides a redundant path for selected CISS measurements to the Centaur vehicle DCU, pulse
code modulation (PCM) system. The CISS DCU provides a PCM stream independent of the Centaur vehicle PCM to the Orbiter PDI, payload recorders, and
the CCLS. The CISS PCM also interleaves the CU data.

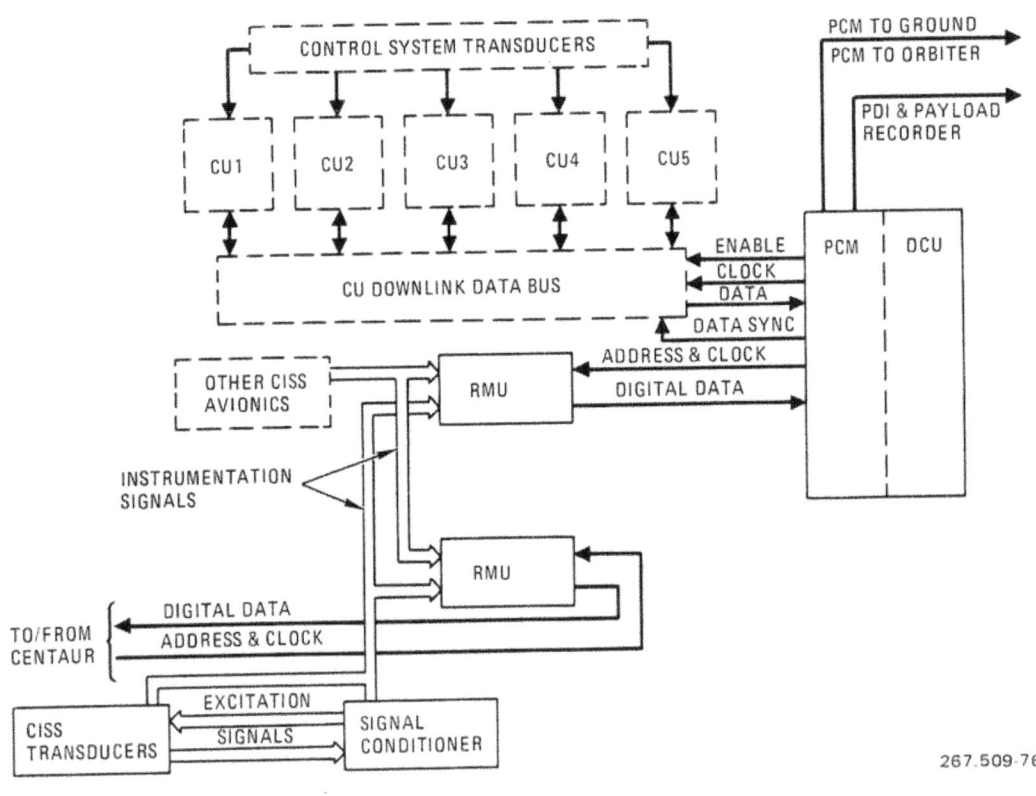

Figure 3-60. CISS Instrumentation System for Centaur G-Prime

d. **Pyrotechnic System.** All Centaur G-prime pyrotechnic functions will be provided by a combination of a pyrotechnic initiator controller (PIC) with a dedicated NASA standard initiator (NSI). The PIC incorporates the necessary safety circuitry and telemetry interfaces. It also provides the necessary energy for sure-fire detonation of the NSI.

e. **Propellant Level Indicating System.** A propellant level indicating unit (PLIU) supplies power to the tanking level sensors and detects propellant level conditions of the sensors. Level indications are provided to the propellant loading control system via the telemetry system and overfill indications are sent to the CUs and also delivered to GSE via TLM.

3.3.5.3 **Orbiter Interfaces.** This system is designed to minimize avionics interfaces with the Orbiter. The interface requirements are met using only elements of the standard Orbiter services with some modifications for system efficiency. Orbiter services used are:

- Telemetry and data services to monitor the status of the CCE in the attached and detached modes.

- Command interfaces for controlling on-orbit deployment functions.

- Displays and controls for crew status monitoring and safing interfaces.

- Power and cable interfaces.

The extent of the Orbiter services required is discussed in the following paragraphs.

a. **Telemetry and Data Services.** The Orbiter interleaves PCM data from one of two spacecraft channels: one Centaur and one CISS. In the attached mode, four individual PDI channels receive serial data and select specific preprogrammed words to be interleaved with Orbiter data transmitted to the ground; 6.4 kilobits of Orbiter downlink bandwidth (including overhead) is available during ascent and 32 kilobits during on-orbit operations. Use of the Orbiter Ku-band system will allow 64 kilobits of CISS and Centaur data to be interleaved through the PDI and sent to the ground.

   In the detached mode, CISS continued to send data to the data interleaver; however, Centaur/spacecraft data will flow to the Orbiter through the payload interrogator. One payload recorder channel and the Orbiter bent-pipe downlink mode supplement the Orbiter's real-time downlink data capability.

b. **Command, Display, and Control Interfaces.** Command inputs from the crew are required for on-orbit deployment functions. Commands are initiated via the standard switch panel or standard software functions controlled by the keyboard. Safety inputs for monitoring will be read through the PDI and/or the payload interrogator in the attached or detached mode and through direct measurements wired into the multiplexer-demultiplexer caution/warning inputs, or from the five channels of the caution/warning electronics assembly.

During predeployment sequences, keyboard commands initiate operations through the payload signal processor to the Centaur DCU only. Commands and feedback responses at the standard switch panel initiate and monitor CISS automated deployment functions. Safety status is brought to the crew's attention by the caution/warning electronic assembly or by data under analysis by standard Orbiter system software. The data will be displayed on cathode ray tubes (CRT). Safety abort dump commands will be initiated either manually with the keyboard or automatically via the GPC with the commander's initiation of Orbiter abort.

c. Electrical Power and Cabling. The CCE uses 28 Vdc power supplied by the Orbiter. Two CISS-mounted silver-zinc batteries are diode coupled to the main dc power bus in the EDU to provide dual-failure tolerant dc power. The batteries will only be used in the event of Orbiter power system failure. No direct ground support cabling is necessary. During ground operations, prelaunch, ascent, and on-orbit, the CCE will use the 3.2 kW provided by the Orbiter as its primary power source.

Cabling that supplies power and all other electrical interfaces consists of four sections of shortened standard mix cargo harness (SMCH) unique for Centaur missions. Cable interfaces afford the CCE the full complement of T-0 umbilical wires allocated for the Orbiter cargo. The combined SMCH capabilities provide the mix of payload services - both avionic and from the flight deck - to meet system requirements at minimum weight. A physical connection is located on the port and starboard sides of the CISS. Attach points are at the forward end of the CISS.

Three sources of 200/115 volt, 400 Hz, four wire ac power are supplied by the Orbiter. The Orbiter supplied ac power is used only in the operation of the deployment adapter rotation system.

3.3.6 SOFTWARE SYSTEM. The Shuttle/Centaur software responsibilities at General Dynamics include the five basic software areas as shown in Figure 3-61.

1. Centaur vehicle software

2. CISS software

3. GCS software

4. CCTE software

5. Support software

The first two elements are considered airborne software since they are hosted in airborne computers (the vehicle DCU, the CISS DCU, and five CUs).

The remaining three are considered ground software since they are hosted only on ground support computers.

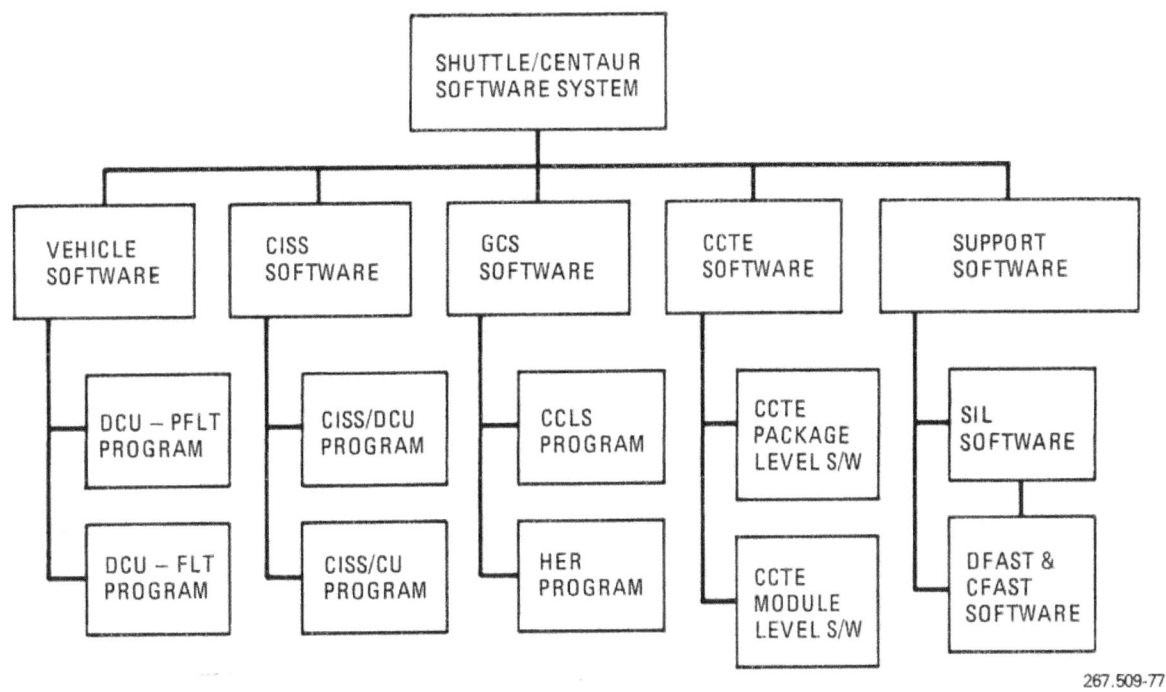

Figure 3-61. Shuttle/Centaur Software System

3.3.6.1 Centaur Vehicle Software

a. Centaur G-Prime DCU Software. Centaur vehicle software supports ground
checkout, launch, and flight operations. It is defined as the software
that executes within the Centaur vehicle DCU.

The software is divided into two basic operational categories: preflight
and flight.

Preflight software covers ground checkout and operations up to Shuttle
launch. Flight software covers all vehicle DCU activities from Shuttle
launch to mission completion.

Safety-related vehicle functions, while Centaur is attached to the
Orbiter, are excluded from DCU responsibility. They are performed by an
independent set of CISS avionics and software. This approach greatly
reduces vehicle software requirements related to single and dual-failure
tolerant operations. The flight DCU software, although operating in a
"single string" avionics system, employs software and hardware backups
that make the flight program operation as forgiving as possible of
critical hardware failures.

DCU preflight software operations are closely integrated and augmented with the GCS ground-launch computer software. This minimizes the software burden on the DCU during ground operations. The preflight software uses storage overlays and limits itself to the 4,096 words of alterable memory to minimize core requirements.

Major functions supported by preflight operations are:

- Centaur and spacecraft systems checkout.

- Centaur IRU calibration and alignment.

- Communications between the Centaur DCU and the ground computers.

- Terminal-count sequencing and testing.

DCU flight software modules are stored in the 12,288 words of DCU non-alterable memory, but also use most of the alterable memory for scratchpad and data storage after vehicle liftoff. Major functions of the flight software are:

- Sequencing

- Guidance

- Navigation

- Control

- Propellant utilization

- Tank-pressure and vent control

- Telemetry formatting

- Communications to and from the Orbiter

- DCU self-test and predeployment tests.

Centaur flight software operates in a passive mode through deployment from the Orbiter. The CISS remains in full control of all vehicle safety-related functions during this period. The only DCU programs operating before deployment, other than the operating system, are:

- Navigation

- Telemetry formatting

- DCU self-test

- System checkout

- Sequencer

- Communication to and from the Orbiter

- Predeployment tests.

The software is structured in a modular concept where each module performs a unique function and represents a convenient and manageable segment of instructions and/or constants individually coded, checked, documented, and maintained in a program library under configuration control. Individual software modules selected from the library are integrated to satisfy the program requirements and undergo integrated program checkout. Table 3-3 (preflight and flight software modules) lists the software modules in each category that are available to support the Centaur G-prime program requirements.

Each DCU module is coded in machine assembly language unique to the DCU instruction set. The library of modules so coded is extensive and flight-proven and includes software programs from Centaur D-1A and Centaur D-1T missions. All modules, because of their machine language coding, are highly efficient and optimized for minimum DCU core usage and/or duty cycle, as the case demands.

Table 3-3. Preflight and Flight Software Modules

Preflight Modules	Flight Modules
Preflight System	Attitude Error and DDR
DCU Alterable Memory Test	Coast Autopilot
DCU RAM Test	Coast Guidance
Ramp Output Generator	Permanent Constants
Sawtooth Output Generator	Centaur Star Scanner
Sine Wave Output Generator	Computer Controlled Vent and Pressure
SCU Switch Test	Flight Initialization
IRU Calibration and Alignment	Powered Guidance
DCU Memory Dump	Hydrazine Monitor
Vehicle End-to-End Steering Test	Navigation
DCU SKD Test	Powered Autopilot
IMG Steering Chain Test	Predeployment Checks
PU Test	Vehicle Telemetry Program
Gimbal Slew Test	Post Injection
DCU Instruction Test	Platform Strapdown Torquing
Combined Function Generator	Propellant Utilization
Terminal Count Sequencing	Discrete Priority
DCU Interrupt Check	Operating System
DCU I/O Driver	Steering Interface
	Mission and Vehicle Peculiar Telemetry
	Vehicle Sequencer
	Antenna Selection
	Navigation Update
	Accelerometer Bias Calibration
	Math Subroutines
	Platform Rotation

b. <u>DCU Resource Summary</u>. Table 3-4 lists the flight program modules and their memory requirements. Previous experience in performing numerous mission-peculiar changes to the DCU flight software for Centaur D-1A and Centaur D-1T programs has developed the expertise, software structure, and techniques to readily and efficiently incorporate a wide class of mission-peculiar changes into the basic Centaur vehicle software.

Peak duty-cycle requirements are estimate to be 80% during coast phases and 88% during powered phase. This spare capacity is considered more than adequate due to the maturity of the baseline software and because the peak is computed from the sum of the individual module peaks, which do not occur simultaneously.

Table 3-4. Estimated Centaur Vehicle DCU Memory Requirements

Flight Module	Centaur G-Prime Core Storage Required
Attitude Error and DDR	150
Coast Autopilot	288
Coast Guidance	400
Permanent Constants and Sequencing Data Table	627
Centaur Star Scanner	750
Computer Controlled Vent and Pressure	889
Flight Initialization	158
Powered Guidance	1,276
Hydrazine Monitor	67
Navigation	470
Powered Autopilot	223
Predeployment Checks	225
Vehicle Telemetry Program	1,721
Post Injection	75
Platform Torquing	591
Platform Rotation	161
Propellant Utilization	270
Discrete Priority	77
Resident Control System	1,550
Steering Interface	598
Telemetry Recovery	150
Vehicle Sequencer	610
Antenna Selection	200
Accelerometer Bias Calibration	150
Slack Task	20

c. Software Documentation and Configuration Control. A software development
plan defines the software development process. This plan is in accordance
with existing software control procedures and NASA Management Instruction,
NMI 2410.6.

Functional requirements document (FRDs) define the software requirements
and serve as the specifications for software development.

The software coding begins after the requirements are documented. The
software is developed, validated, and controlled initially on a
module-level basis, and expanded to the program level as the module
checkout is completed.

The major software configuration control tool is the check figure, a
number generated by adding the computer instructions within a module or
program. The check-figure changes not only when instructions within a
module change, but also when the sequence of the instructions changes.
Source-code check figures are used to control the module source programs,
object-code check figures for the object code, and total-load check
figures to control the entire DCU memory load. Computer load is inhibited
in the event of a check-figure discrepancy. The check-figure control is
used for both the flight program and constants load tapes.

d. Deployment Delays. The Centaur DCU is synchronized to real-time (GMT)
through the GSE before liftoff. Since the system is navigating from the
ground up, the Centaur is insensitive to delays in deployment of the
Centaur from the Orbiter. This permits the mission to be accomplished on
any deployment opportunity. For those missions that have a time-dependent
target condition (such as the hyperbolic asymptote), the appropriate
target-orbit data is computed within the DCU once the deployment time is
established. This feature is also a part of the baseline Centaur G-prime
software in which the planetary targeting is performed onboard the DCU
after deployment, using prestored trajectory data (in this case, the
desired hyperbolic asymptote and orbital energy). The synchronization of
DCU time and GMT thus makes the Centaur system autonomous, and obviates
the need for crew-initiated functions to initialize the software or
communicate the deployment time.

e. Rapid Targeting. The baseline guidance equations are a simple modification of the existing Centaur D-1A and Centaur D-1T guidance equations (the
modifications are primarily the deletion of D-1A and D-1T peculiar
functions). These equations are an explicit formulation of the two-point
boundary value problem and, as such, require a minimum of targeting data
and no a priori trajectory shaping information. This permits the
targeting process to be extremely simple and enables the target orbit to
be selected very close to launch.

The guidance software is validated over the entire mission opportunity, using representative target orbits. The validation is intended to test the guidance equations and airborne implementation using nominal, 3σ and abnormal ($>3\sigma$) environments, but not necessarily the particular guidance constants of a specific mission. Targeting, which is the generation of the target orbit guidance constants, proceeds as follows:

1. The target orbit parameters are selected anywhere within the range of the previously validated mission opportunity. The target parameters can include apogee, perigee, eccentricity, argument of perigee, inclination, and ascending node (geocentric missions), or asymptote and energy (planetary missions).

2. The selected parameters are input to the closed-loop guided simulation and a set of test cases (nominal and 3σ environments) is executed in the general purpose, nonreal-time simulation to verify targeting accuracy.

3. A constants tape is written with the parameters of the target orbit, and firing tables are generated. (Firing tables aid launch site personnnel to verify the contents of the flight constants tape.)

4. The constants tape and firing tables are transmitted to the launch site.

The current targeting ten weeks before launch of Centaur D-1A FLTSATCOM and INTELSAT missions is typically accomplished in less than five days. Since there is no design work in this targeting process, it can be automated and carried out comfortably 48 hours before launch.

f. Software Validation. The analysis and validation of the flight software is performed on three levels and with three simulation tools. The three levels of testing and validation are:

1. Design, analysis, and evaluation of the flight algorithms.

2. Module-level checkout of the flight software.

3. Integrated systems level checkout of the flight software.

The overall process is illustrated in Figure 3-62. The design, analysis, and evaluation of the flight algorithms (e.g., guidance, navigation, etc.) is performed on the Cyber 760 general-purpose computer, using Fortran replicas of the flight software in a nonreal-time mode. The software is tested using nominal, 3σ, $>3\sigma$, and failure-mode environments for the complete range of target conditions, i.e., the complete mission opportunity. Compliance with software accuracy requirements is established with these tests, except for inaccuracies caused by the flight-computer truncation.

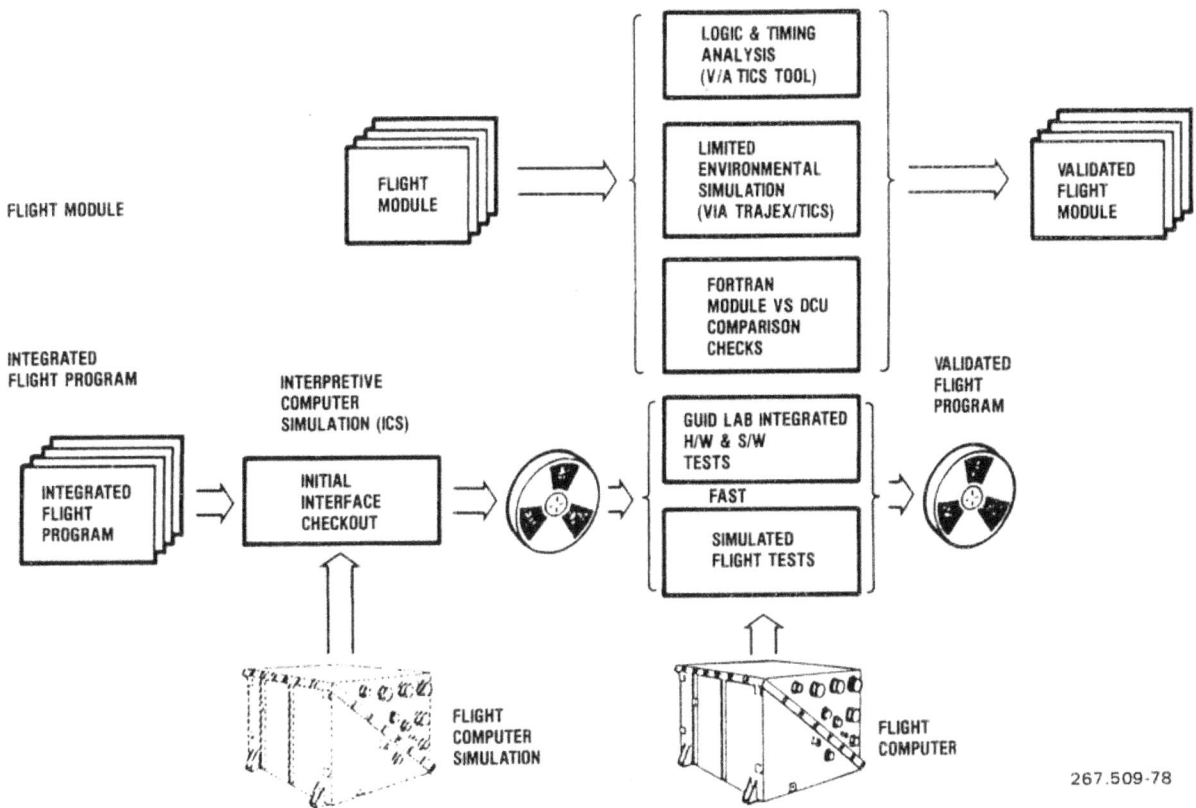

Figure 3-62. Centaur DCU Software Checkout

Module-level tests are performed using the interpretive computer simulation (ICS). The ICS is discussed in Section 3.3.6.5. These tests are designed to validate the airborne code of each module in the flight program. The outputs from these module-level tests are compared to the outputs from the Fortran replicas driven by identical inputs. This establishes truncation errors due to the flight DCU. The ICS also provides scaling evaluation of the fixed-point DCU code, and timing analysis of each module.

Integrated software checkout is also performed with the Cyber ICS. In this mode, the ICS is run closed-loop in conjunction with the general-purpose trajectory simulation and executes all, or subsets, of the flight modules. This simulation validates interfaces between the software modules, and provides timing data at the integrated systems level. The ICS in this mode is nonreal-time.

The integrated systems-level validation is performed in the SIL with the actual-flight DCU code executing in a flight configuration DCU, in real time, in a closed-loop guidance mode in conjunction with a general-purpose vehicle simulation. This simulation, the flight analogous software test DFAST, is discussed in Section 3.3.6.5, and provides all the external I/O to the flight program. The DFAST tests provide a complete system-level software interface validation, and final evaluation of DCU truncation and timing. The output from this test is orbital data, analog traces of significant DCU parameters, and a telemetry tape of all DCU data for post-processing analysis.

3.3.6.2 <u>CISS Software</u>. The CISS software executes within the five CISS CUs plus the CISS DCU. The major purpose of the CISS software is to control, in a one- or two-failure tolerant system, all safety-related functions while the Centaur vehicle is attached to the Orbiter. This design relieves the burden which would otherwise be placed on the Centaur vehicle avionics to control these functions in a two-failure tolerant mode. The CISS software supports:

● Ground testing at the factory, Complex 36, Vertical Processing Facility, and Complex 39.

● Ground and launch support operations, including Centaur vehicle tanking.

● Testing and operations of Centaur and CISS through separation.

● Prelanding and postlanding operations, including abort.

During ground operations the software is closely monitored and controlled by the GCS. The GCS can communicate to any of the five CISS CUs or CISS DCU. After liftoff, the CISS software is controlled via internal CU software sequencing and via discrete commands from the Orbiter.

a. <u>CISS Control Unit Software</u>. The basic software design approach for the CUs is similar to that employed for the Centaur DCU. The software uses the same modular concept and is operationally controlled by a software task scheduler. The CUs are coded in a machine assembly language. The software is being developed on the General Dynamics Software Engineering System (SES). It is documented and under configuration control similar to that for Centaur DCU software.

Table 3-5 lists the CU functional control modules for each of the five CUs. A minimum amount of system checkout and CU self-testing is planned. This simplified test approach is made possible by the avionics systems design, whereby control of a system is at the functional level and the five CU outputs are voted at the relay level. These concepts greatly simplify both the magnitude and complexity of the software, which otherwise would have to address itself to the problems of fault detection and system reconfiguration. The memory for each CU is 16K of 8-bit words PROM and 4K of 8-bit words RAM.

The CU software design has several significant features.

● Each CU operates independently to avoid single-point failures.

● Time synchronization of each CU's functional program operation is not required.

● Downlink data from each CU is synchronized at the start of a data cycle to avoid double-decommutation of PCM data.

● Software in the programmable read only memory (PROM) of all CUs is identical to maintain CU interchangeability.

Table 3-5. Control Unit Functional Control Modules

FUNCTION	
Computer Control System	Functional Control System
Telemetry	Sequencer
Task Scheduler	Tanking (LH_2, LO_2)
I/O	Testing
Exec. Control	Press. & Vent Control
Uplink Control	ΔP Control
Interrupt Logic	Dump Logic
Initialization	Power Monitor & Control
	Purge
	Pneumatics
	Deployment
	Helium Supply

- CISS CUs actively control all safety-related Centaur vehicle operations up through deployment; e.g.,

  - Sequencing

  - Pressurization and vent control

  - Abort operations

  - Purge system control

  - Pneumatic system control

  - Deployment operations

  - Pyrotechnics

- GCS will be able to communicate independently with each CU.

- CUs will not be required to communicate with one another or communicate with the Centaur DCU.

b. CISS DCU Software. As its main function for both preflight and flight operations, the CISS DCU will downlink PCM data gathered from the CISS RMU and the five CUs. The only software required is an operating system similar to that used by the Centaur vehicle DCU and one or more unique PCM formats to satisfy data downlink requirements.

3.3.6.3 GCS Software. GCS software supports vehicle checkout and verifies the launch readiness of the Centaur avionics. Checkout is performed by communicating test instructions to the Centaur DCU, which in turn provides test stimuli to the vehicle avionics. GCS, in turn, monitors vehicle responses via PCM telemetry and verifies proper conditions and tolerances.

The basic requirements for GCS software are:

- Control and monitor Centaur and CISS avionics systems
- Control and monitor Centaur tanking operations
- Adequately exercise systems for thorough problem isolation
- Maintain flexible operator command and control
- Generate accurate historical data records
- Perform testing safely with minimum chance for human error.

The GCS software is divided into modules which perform system and execute functions and applications programs that perform validation, airborne testing, and vehicle functional operations. All GCS software is documented, released, and under configuration control.

3.3.6.4 <u>Computer Controlled Test Equipment (CCTE) Software</u>. CCTE software is composed of three software groups to allow checkout and evaluation of Centaur avionic packages and their associated cards and modules. These three software groups are:

- The circuit card and module software associated with the HP1000 computer to checkout analog boards. These programs are written in Fortran.

- The circuit card and module software associated with the HP1000 computer to checkout digital (TTL) boards. These programs are written in Test Aid III and Fortran.

- The package-level checkout software associated with the H100 computer to perform functional, temperature, and vibration package testing. These programs are written in Fortran.

3.3.6.5 <u>Support Software</u>. The Centaur/CISS DCU support software, which is mission independent, consists of the following elements:

1. Interpretive computer simulation (ICS)
2. Flight analogous software test (DFAST) integrated system simulation
3. Assembler
4. Loader
5. Librarian.

These tools are derived from the existing Centaur D-1A program and are being adapted to the Centaur G-prime program with only minor revisions. The ICS is being modified to add additional I/O functions consistent with computer modifications. The DFAST simulation is being augmented with a Shuttle ascent simulation and platform torquing functions, and is being translated from a Harris/4 to a Harris/H800 computer. There are no changes to the existing assembler, loader, and librarian.

The ICS is an exact simulation of the DCU instruction set and interrupt structure. Bit-for-bit agreement has been verified between the simulation and the actual DCU hardware for all operations. The simulation contains diagnostic aids to permit the user to assess scaling of all variables, and timing of any subset of instructions, any module, or the entire integrated program. The ICS is typically used at both the module level and at the integrated system level.

For module-level checkout, special-purpose test drivers are created to execute the module under test. At the integrated-system level, the general-purpose trajectory simulation is used as the executive program, and supplies the simulated external input to the ICS, which in turn simulates the DCU. All I/O is executed in this mode, including IRU outputs, tank pressures, and propellant sensors. In turn, the ICS outputs steering commands to the trajectory simulation to close the guidance and control loop. The ICS is resident on the Cyber computer and has been used for ten years on the Centaur D-1A and Centaur D-1T programs.

The DFAST simulation is a simplified six-DOF simulation of the launch vehicle system that operates in real time and interfaces with a DCU loaded with a flight program. The DFAST simulation is resident on the Harris/H800 computer in the SIL. This facility is used for extensive system-level validation of the integrated flight program, and exercises the actual flight-program code in both test and flight DCUs. The DFAST simulation is used primarily to validate the operation of the integrated software, the external interfaces to the DCU, and the internal software interfaces in a realistic flight environment. This includes nominal, abnormal, and failure-mode environments. All external I/O and interrupt functions are exercised by the DFAST simulation. These include:

1. Analog to digital inputs to the DCU

2. Digital to analog outputs from the DCU

3. Power on, power off, real-time (50 Hz), GSE, and telemetry interrupts

4. Discretes input to the DCU

5. Discretes output from the DCU

6. Orbiter uplink commands and data.

The DFAST simulation outputs include trajectory and vehicle-state data, analog traces of key DCU variables, and a DCU telemetry tape. The telemetry tape is processed with a companion post-processing program.

The CU Flight Analogy Software Test (CFAST) simulation is a real-time CU environment simulation which is used for testing CU flight softwre. CFAST is hosted on the software development computer system Harris/H800 computer in the SIL and simulates the following CU interfaces:

1. Power supply

2. Discrete inputs

3. Analog inputs

4. Uplink

5. Downlink

6. Caution and warning hardline (Happy Flag)

7. Relay matrix

CFAST simulates the fluid systems, tank pressures, and rotation system which the CU software is designed to control, and is the primary tool for CU software integrated system validation.

CFAST simulates the CU environment from Shuttle launch to landing. An inflight-restart capability allows the user to stop a simulation at selected points and save the simulation and CU data required to continue. At a later time, the user may restart the simulation using these data sets. Failures and non-nominal conditions may be introduced into the simulated control subsystem models.

CFAST has the capability to load flight software into all-RAM test configuration CUs during the development and testing process.

Output is provided on stripchart recorders, line printers, magnetic tapes, and teletype error messages.

The assembler for the DCU translates symbolically coded instructions (source program) into a relocatable, machine-oriented language. This assembler output is termed the object program. The source programs input to the assembler are maintained and modified under the CDC update utility. Use of the update utility provides a record or "tracks" of all changes made to a program. Finally, the assembler produces the source check figure and the loader check figure.

The loader for the DCU is a linking loader that takes relocatable object programs as input and produces an absolute load file for the target computer. The loader computes the check figure for each program module and compares it with the check figure computed by the assembler. If a miscompare is found, a diagnostic is printed and the load is terminated, thus ensuring that loader reads the object program correctly. When the load is completed, the loader computes a core-load check figure for the absolute load. This core-load check figure is sensitive to any change in the program, including the order in which the program modules are loaded. The core-load check figure then becomes the configuration control tool used to identify the program tape used at DCU load time.

The librarian for the DCU software provides a facility for maintaining a library of object programs (assembler output) and their assembly listings. The librarian allows the user to select, by program module name, any mixture of previously assembled object programs. The librarian builds a file of object programs for input to the loader and a printed output assembly listing of each program selected. The librarian thus provides a cost-effective method of building a flight program without reassembling each module, and it assists configuration control by selecting known, previously tested, object program modules.

The assembler, loader, and librarian are all resident on the Cyber computer.

The support software for the CUs is being developed on the SES. It uses the Harris Vulcan operating system, a multiuser, priority-structured memory-management program which concurrently maintains multistream batch processing, interactive time-sharing, database management, remote job-entry, and real-time operations.

The support tools available with this system and used for Centaur G-prime software development include:

● Z-80 microprocessor cross-assembler and compiler

● Zilog Z-80 simulator

● Software product management system

● Data analysis

● Software library system

● Automatic flow charting

● Documentation

● Graphic display

3.3.7 MASS PROPERTIES. Mass properties presented in this section are consistent with performance data in Section 3.2, Baseline Mission.

3.3.7.1 Centaur/Spacecraft Weights Summary. A weight summary for Centaur G-prime with the Galileo spacecraft is given in Section 3.2.3, Table 3-2.

3.3.7.2 Centaur Airborne Support Equipment Weights. Table 3-6 lists weights of individual major components of Centaur G-prime ASE.

Table 3-6. Centaur Airborne Support Equipment Sum

Item	Weight (lb)
Total Airborne Support Equipment (TASE)	9572
Centaur Integrated Support System (CISS)	6672
Centaur Support Structure (CSS)	2367
Deployment Adapter	580
Fluid Systems	1495
Avionics	1402
Rotation System	322
Separation System	180
Spacecraft Required LVMP	148
Residuals	178
Orbiter Mounted Items	2900

SECTION 4

SYSTEM REQUIREMENTS

4.1 INTEGRATION REQUIREMENTS

Integration requirements for Centaur G-prime are based upon the Space Shuttle
and mission requirements.

Centaur G-prime is integrated into the Space Transportation System through the
efforts of the various integration panels and working groups at the NASA
Johnson Space Center, Kennedy Space Center, and Lewis Research Center. The
panels are reviewing and controlling the integration of Shuttle-to-Centaur
interfaces as well as handling any spacecraft-peculiar requirements. The
various interfaces are being worked in conjunction with Rockwell and JSC to
ensure complete compatibility with the Space Shuttle Orbiter.

The Centaur G-prime stage is integrated into Shuttle to minimize the impact on
Shuttle. Centaur G-prime will use the standard Shuttle interfaces, as defined
in ICD 2-1F001, "Orbiter Vehicle/Centaur".

The Centaur cargo element is designed to keep the Centaur vehicle as
independent as possible of the Shuttle system or operation. The Centaur
integrated support system (CISS) is used to provide interface compatibility
between Centaur and the Orbiter. In addition, the CISS controls all
safety-critical functions for the Centaur vehicle with the level of redundancy
necessary for the identified hazards.

Servicing of the liquid hydrogen propellant will be through the port-side
midbody T-0 panel; liquid oxygen will be serviced through the aft starboard
T-0 panel. If mission abort is required, the Centaur propellants may be
dumped any time before Super*Zip firing for Centaur separation. This will
allow a safe landing for the crew and Orbiter. Propellants will be dumped
upon command from either the Orbiter computers or from the crew. These
activities are integrated into Orbiter operations and timelines.

Centaur operations are integrated into the Shuttle crew timelines and
operations. The goal is to give the crew control of initiating such critical
functions as rotation, separation, or abort.

The thermal environment for the entire Centaur cargo element with spacecraft
is being analyzed with the Orbiter to ensure temperature compatibility. Air
conditioning and purging requirements are being modified or defined.
Propellant tank heating rates are being analyzed under the most severe thermal
environments to which Centaur will be exposed to ensure pressure control in
both tanks.

An analysis of the interface loads will be based on the load factors given in ICD 2-1F001. A dynamic analysis using the Shuttle model and forcing functions, as obtained from Rockwell, will be used to verify acceptability of the interface loads.

During prelaunch activities at the Eastern Launch Site, Centaur will be mated to the CISS at Complex 36 where it will be checked as an element before integration and test with the spacecraft and Shuttle. The Centaur/spacecraft will be installed in the Shuttle in the vertical mode. They will be capable of being removed horizontally while the Orbiter is in the Orbiter Processing Facility following a flight abort.

Wherever possible, systems designs and interfaces will be implemented using existing flight components and ground support equipment.

4.2 SYSTEM SAFETY

The system safety program for Shuttle/Centaur ensures that Centaur will generate no hazards that endanger the Space Transportation System.

This level of safety will be achieved by incorporating the system safety engineering discipline into the detailed design, assembly, test, and ground and flight operations of Shuttle/Centaur.

System safety requirements of NHB 1700.7A will be met.

The hazardous aspects of operating a cryogenic upper stage in the STS have been under analysis since the initial Space Tug studies of 1972. These analyses have determined the Centaur can safely operate as an STS element. As detailed safety requirements have been identified, the design of the Centaur D-1A has evolved into the Centaur G-prime to satisfy them.

Early studies led directly to the 1982 Phase 1 safety review to the requirements of NHB 1700.7A. The Phase 1 Review of the Centaur vehicle and airborne and ground support equipment was conducted by both the JSC and KSC Safety Committees. Subsequent safety reviews of the Spacecraft Integration and Centaur Fluid System changes have been held. They concluded that Centaur can safely operate from the Orbiter payload bay.

The integrity of the propellant tanks to contain cryogenic fluids and the validity of the tank analyses have been demonstrated over the last 20 years with launches using both D-1A and D-1T boosters.

Redundancy to provide the safety necessary for Centaur G-prime has been provided in the fluids, mechanisms, and CISS avionics systems. Appropriate parallel and series valves have been added to the tank pressurization, vent, and drain/dump functions to ensure a single valve failure does not lead to a hazard. Mechanisms for erecting the Centaur G-prime are also redundant to achieve the required failure tolerance.

The avionics controlling and inhibiting safety critical events have been designed to be fully independent of the Centaur guidance, navigation, and control system. This separation allows flight functions to be accomplished by the fault-tolerant CISS system before Centaur/Orbiter separation. No matter what the condition of the basic Centaur avionics, the CISS will retain control and prevent the occurrence of hazards. CISS control functions are implemented by five control units that allow three out of five, or two out of three voting to accomplish the required control redundancy. The Centaur DUFTAS unit is used to provide fault-tolerant control and inhibiting of hazardous functions normally occurring during Centaur flight.

The system safety program will continue to build on our safety studies. Hazard reports will continue to be generated as hazards are identified and closed out when, with management and customer concurrence, appropriate preventive measures are defined. Hazard identification will be accomplished using failure modes and effect analysis, engineering analysis, and specific safety analysis. Periodic reviews of the hazard reports by NASA will be accomplished for the Phase II and III Safety Reviews.

System safety engineers assigned to the Shuttle/Centaur program will participate in all trade studies performed on the Centaur and its support systems. They will also participate in all design and test reviews, both in-house and with the customer.

4.3 RELIABILITY AND QUALITY ASSURANCE REQUIREMENTS

Twenty years of reliability and quality growth have led to a mature operational system for launch vehicles.

The Centaur G-prime reliability/quality program meets the intent of NHB 5300.4 (1D-2), as defined in the Mission Assurance Plan, Convair Report BGJ 72-006. Reliability and quality are designed into GDC products and ensure there is no degradation of inherent design reliability through the succeeding steps from fabrication to end use.

During the design phase, functional and environmental requirements are translated into functional requirements documents. Safety margins, derating factors, and failure effects will be developed and analyzed. Physical parameters and constraints will be addressed and test requirements, including overstress tests and test quantities, will be identified.

Reliability effort includes failure modes and effects analyses (FMEA) and a critical items list (CIL). These efforts emphasize the identification of single-point failures at the system and subsystem level to determine possible modes of failure and their effects on mission objectives and crew safety.

A complete electronic parts control program will be implemented through the Space Parts Control Board, as defined in the Mission Assurance Plan, including parts selection in the order of precedence established by MIL-STD-143B. This includes:

- Space Approved Parts List (SPAPL), GDC Report 55-20777

- MIL-STD-975C

The parts program also includes a derating policy, qualification of piece-parts, and detailed control drawings.

Quality tasks are identified in the Mission Assurance Plan. These tasks include:

- Design assurance
- Process control
- Identification and data retrieval
- Procurement control
- Fabrication control
- Inspection
- Nonconformance control.

All of these tasks are being accomplished for the Centaur D-1A program and will be continued for the Centaur G-prime.

SECTION 5

MECHANICAL FACILITIES AND GROUND SUPPORT EQUIPMENT

5.1 GENERAL CONCEPT

Integrating Centaur into the Space Transportation System requires modifications to certain Convair Division, Cape Canaveral Air Force Station, and Kennedy Space Center (KSC) facilities and some new and modified ground support equipment (GSE). These changes are required for one or more of the following reasons.

- To allow for tank geometry differences between the Centaur G-prime and the Centaur D-1A configuration.

- To accommodate the Centaur integrated support system (CISS) and to facilitate its assembly with Centaur.

- To simulate certain aspects of the Orbiter/Complex 39 installation at Complex 36A.

- To provide the necessary portable equipment to service the Centaur and CISS assembly (CCA) while at KSC (Figures 5-1 and 5-2).

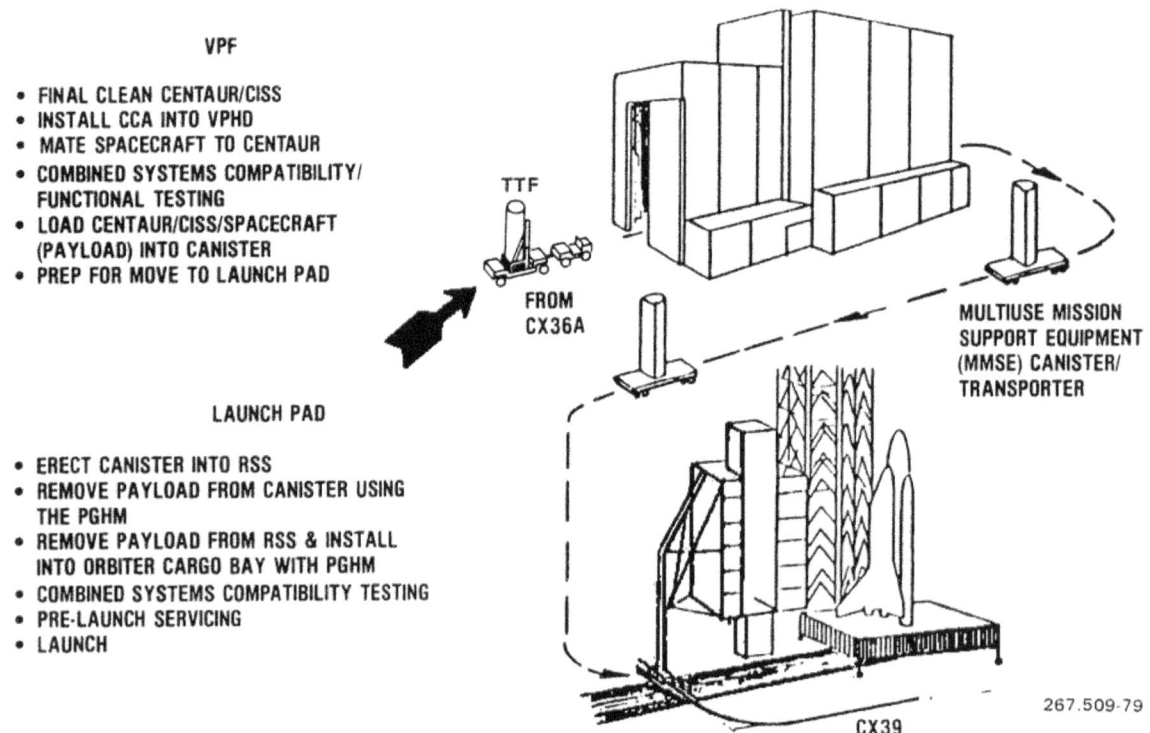

VPF

- FINAL CLEAN CENTAUR/CISS
- INSTALL CCA INTO VPHD
- MATE SPACECRAFT TO CENTAUR
- COMBINED SYSTEMS COMPATIBILITY/ FUNCTIONAL TESTING
- LOAD CENTAUR/CISS/SPACECRAFT (PAYLOAD) INTO CANISTER
- PREP FOR MOVE TO LAUNCH PAD

TTF

FROM CX36A

MULTIUSE MISSION SUPPORT EQUIPMENT (MMSE) CANISTER/ TRANSPORTER

LAUNCH PAD

- ERECT CANISTER INTO RSS
- REMOVE PAYLOAD FROM CANISTER USING THE PGHM
- REMOVE PAYLOAD FROM RSS & INSTALL INTO ORBITER CARGO BAY WITH PGHM
- COMBINED SYSTEMS COMPATIBILITY TESTING
- PRE-LAUNCH SERVICING
- LAUNCH

CX39

267.509-79

Figure 5-1. Prelaunch Operations

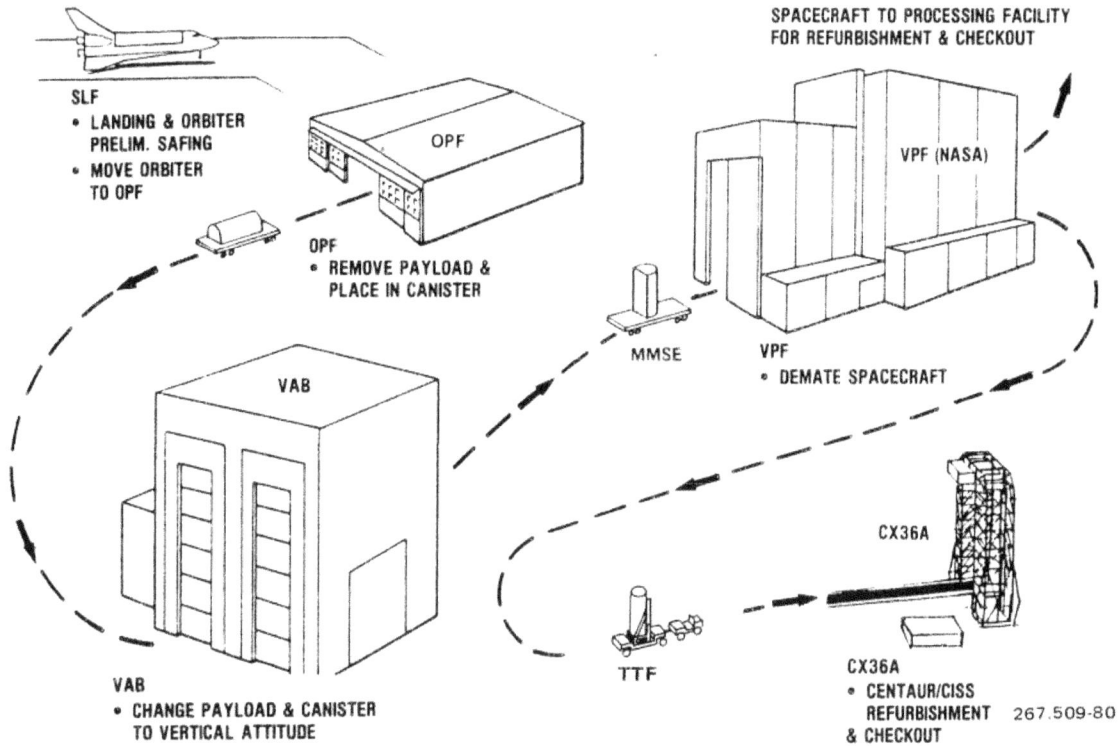

Figure 5-2. Centaur G-Prime Abort Flow Path

5.2 GENERAL DYNAMICS CONVAIR (GDC) FACILITIES, SAN DIEGO

Changes to General Dynamics facilities in San Diego are primarily modifications to accommodate the change in Centaur tank geometry. Fabrication of all versions of the Centaur tank will be performed at Air Force Plant 19. The Centaur G-prime will be transported to the GDC's Kearny Mesa plant on a modified Centaur D-1A transport pallet.

CISS fabrication and all subsystem installation and checkout will be performed primarily in Building 5 of the Kearny Mesa plant. The dock area has been modified to accommodate the new tank configuration. Changes to existing pneumatic checkout equipment for testing the Centaur and CISS will be minimal. The CISS will be supported in the test and transport fixture (TTF) during checkout and transportation. The vertical assembly tower (VAT) will be modified to allow vertical assembly and checkout of the Centaur and CISS supported in the TTF.

5.3 CAPE CANAVERAL AIR FORCE STATION (CCAFS) FACILITIES

The most significant change to the CCAFS facilities will be at Complex 36A. It will be modified to allow assembly of the Centaur, CISS, and insulation system, and to check them out after assembly. The Centaur D-1A launcher will remain in place and an adapter will be installed to hold the Centaur CISS assembly (CCA) in the vertical position using the TTF. An air conditioned environmental enclosure will be provided by the TTF. Where feasible, the Orbiter/Complex 39 installation will be simulated. The TTF will simulate the Orbiter payload bay nitrogen purge. The transfer lines connecting the LO_2 control skid, LH_2 control skid, and helium control skid to the CISS, and portions of the Centaur LO_2 and LH_2 tank ground vent systems will also simulate the Cx 39 installation.

5.4 KENNEDY SPACE CENTER (KSC) FACILITIES

At the Vertical Processing Facility, Rotating Service Structure, and Orbiter Processing Facility, the Centaur/CISS assembly will use existing KSC handling equipment. GDC will provide the required slings and handling adapters. Access will be provided by existing work platforms with small portable workstands provided by GDC where necessary. Interfacing the Centaur and CISS assembly with the facility pneumatic system will be accomplished with hoses provided, as required, by GDC. Transport between the KSC facilities will be in the multi-use mission support equipment (MMSE) canister provided by KSC. Following a normal mission, the CISS will be removed from the Orbiter in the Orbiter Processing Facility (OPF) using the OPF bridge crane strongback, and placed in the test and transport fixture mounted on the Centaur/CISS transporter (CCT) provided by GDC for return to Hangar J.

Centaur propellants will be loaded and the airborne helium bottles charged during launch countdown at Complex 39. All fluids will come from Shuttle supply sources and will be controlled by the LO_2, LH_2, and helium control skids. Piping required to interface the skids with the Orbiter and supply sources will be provided by KSC. Hydrazine will be loaded at Complex 36A before CCA departure for KSC.

5.5 GROUND SUPPORT EQUIPMENT

Mechanical GSE required to integrate the Centaur with the Orbiter consists primarily of structural and fluid control items. The major structural items include the Centaur transport pallet, the CISS transport pallet, test and transport fixture, Centaur/CISS transporter, and various slings required to handle major airborne and ground elements.

Centaur G-prime will be transported from San Diego to the Cape Canaveral Air Force Station via the NASA Super Guppy.

Fluid control items required to support the Centaur G-prime include the LO_2, LH_2, and helium control skids plus the standby pneumatic control unit. Control skids will be used for propellant transfer operations at Complex 36A and launch operations at Complex 39. The standby pneumatic control unit maintains Centaur tank pressures after installation of the Centaur into the CISS whenever the airborne pressurization system is not in control.

SECTION 6

ELECTRICAL GROUND SYSTEMS AND EQUIPMENT

The ground computer system (GCS), launch control, ground telemetry system, and landline instrumentation systems at Complex 36A are an update of the current system used to support Centaur D-1A. A GCS is comprised of a primary and a backup computer controlled launch set and a hardware extension remote (HER) containing a primary and backup remote interface controller (RIC). GCS, working in conjunction with other ground support equipment, will control the vehicle avionics, the CISS avionics, and the tanking and pressurization skids.

The monitor and control interface with the vehicle will be via mobile or fixed support equipment (MSE/FSE), which will contain the HER, landline instrumentation, and remote launch control equipment. The MSE will be transportable and capable of supporting operations at Complex 36A, or the Vertical Processing Facility (VPF). Identical fixed support equipment (FSE) will be hardmounted at Complex 39A and 39B. Communications between the Complex 36A equipment and the MSE/FSE will be via existing communication transmission networks located throughout the Eastern Launch Site (ELS). Figure 6-1 shows the Shuttle/Centaur electrical ground system.

6.1 GENERAL DYNAMICS FACILITIES

Electrical ground systems are duplicated in San Diego for software development, simulation, and interface verification.

6.1.1 COMPUTER-CONTROLLED LAUNCH SET. Two CCLS systems will be located at GDC in San Diego and will be used in conjunction with an MSE and ground station to provide a duplicate of the planned system at the Eastern Launch Site. The San Diego CCLS will monitor and control Centaur vehicle and CISS operations in the factory and will be used to develop and test all Centaur and CISS computer software. The flight analogous software test (FAST) system will be used for software development and software acceptance testing for all Centaur and CISS computer software used after T-O (launch). The preflight analogous software test (DFAST) will be used for software development and acceptance testing for all Centaur computer software used before T-O (prelaunch) and after Orbiter landing (postflight).

6.1.2 HARDWARE EXTENSION REMOTE. One HER system will be located at GDC in San Diego and will be used to provide a remote interface to the CCLS, which allows the CCLS to communicate with the Centaur and CISS flight hardware, associated GSE, or their simulators.

6.1.3 SYSTEM INTEGRATION LABORATORY. The Shuttle/Centaur Systems Integration Laboratory (SIL) is used to verify adequacy of operational interfaces with the Orbiter aft flight deck and the spacecraft. It is also used to test the mobile support equipment, CISS, and Centaur avionics interface designs. In addition, the SIL will be used for CCLS, DCU, and CU software development and

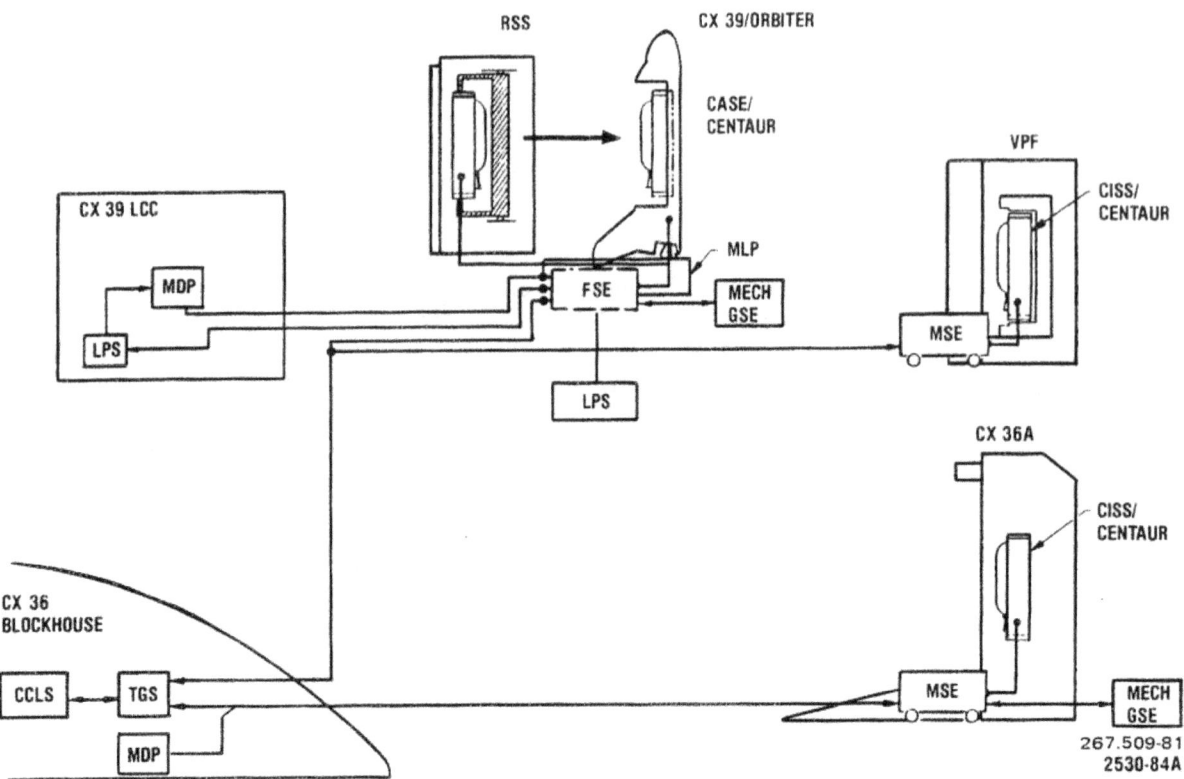

Figure 6-1. Ground Systems Overview

hardware/software integration testing. Actual flight hardware or emulated
electrical equivalents will provide simulation of Centaur, CISS, spacecraft,
Orbiter, deployment, and tanking interfaces. The SIL is modular to enable use
of the simulators at the ELS to verify interfaces during development of the
ELS facilities.

6.1.4 REAL-TIME SIMULATION LABORATORY (RTSL). The RTSL uses an Interdata
8/32 simulation to verify integrity of the CISS control system.

6.1.5 SOFTWARE ENGINEERING SYSTEM (SES). The SES is a Harris H800-based
software engineering system. It will be used to develop the microprocessor
software for the CISS control units.

6.1.6 COMPUTER-CONTROLLED TEST EQUIPMENT (CCTE). The CCTE is a computer-
based automatic test set employed for checkout and evaluation of Centaur
avionic packages and associated circuit cards and modules. The CCTE system
will be used for production testing, and to control and monitor initial and
final acceptance tests, including temperature and vibration tests. All
packages will be checked out by the CCTE in a consistent "hands-off" manner
with appropriate history documentation traceable from component testing to
launch.

6.2 EASTERN LAUNCH SITE FACILITIES

Electrical ground systems at ELS provide vehicle command, control, feedback, and monitoring of Centaur airborne systems.

6.2.1 GROUND COMPUTER SYSTEM. Two CCLS systems will be used to check out and launch the Shuttle/Centaur. Each CCLS consists of a general-purpose computer with standard peripherals, a standard disc-based operating system, and General Dynamics-developed test programs. Each computer will interface to a local digital interface electronics (LDIE) and two operator consoles.

The avionics console will control and test the Centaur and CISS avionics systems. The fluids console will control all fluids and tanking operations.

Commands to the vehicle will be sent via the LDIE, which contains long-distance serial bus protocol electronics (up to 25 miles), precision clock, time-code translator, and high-speed parallel PCM DMA bus electronics. The long-distance receiver will be the HER and will be located in the MSE/FSE.

The second CCLS system at Complex 36 will be in a standby-backup configuration, whereby the operator can switch to the backup system within five minutes of a detected failure of the prime unit.

6.2.2 HARDWARE EXTENSION REMOTE. A HER system will be located in each MSE/FSE at the Eastern Launch Site and will be used in conjunction with CCLS, to control and monitor the vehicle avionics, the CISS avionics, and the tanking and pressurization skids.

Each HER will contain two remote interface controllers (RIC) with independent control of remote functions. When placed in the remote mode, each RIC is tied to the CCLS system. The dual-RIC configuration operates under a watchdog-timer configuration. If one RIC fails, the other RIC is switched in to maintain fail-safe operations.

6.2.3 LAUNCH CONTROL FUNCTIONS. The monitor and control interface with the vehicle will be via the MSE/FSE. The MSE contains the HER, landline signal conditioning and multiplexer, FM electronics, and launch control relay chassis. The MSE supports operations at all locations from initial testing through launch, except at Cx 39 which is FSE.

Manual detanking panels provided by KSC located in the Complex 39 LCC and Cx 36 blockhouse are hardwired into the FSE to permit draining vehicle propellants in a safe and timely manner should circumstances so dictate.

6.2.4 LANDLINE INSTRUMENTATION (LLI) SYSTEM. The LLI system as presently configured for Complex 36 will be used as much as possible with minimum modification by using existing recording systems, signal conditioning systems, patch panels, transducers, J-boxes, cabling, etc.

To support the Centaur, CISS, and associated GSE, the following additions to the Complex 36A landline system are required:

1. Transducers, cabling, and J-boxes for the new LO_2, LH_2, and helium skids.

2. A multiplexer (MUX), signal conditioning system, J-boxes, and cabling in the MSE/FSE to monitor propellant skid hardware and MSE originated measurements. The MSE/FSE pulse code modulated (PCM) data is transmitted via RF hardline to the Complex 36 ground telemetry station for stripout and display.

6.2.5 <u>GROUND TELEMETRY SYSTEM</u>. The ground telemetry system (GTS) provides input/output interfacing with conditioned data/communications transmission lines, recording of PCM signals in a serial digital form, and conditioning, conversion, and decommutation of PCM data. The GTS located at Complex 36 will be upgraded to handle the Centaur G-prime PCM requirements. PCM data decommutators, DACs, and discrete data latch/drivers will be provided to support data displays and CCLS requirements.

6.2.6 <u>TELEVISION (TV) SYSTEMS</u>. The TV system permits remote observation of the tanking operations at Complex 36 and 39. Six new TV cameras will be mounted in the TTF to supplement the existing Complex 36 TV system. Complex 39 TV cameras will be strategically placed and transmitted to Complex 36 over the existing wideband system.

6.2.7 <u>RF SYSTEMS</u>. The RF system will provide for reradiation of the Centaur S-band telemetry signal from Cx 36A, VPF, or Cx 39. The systems will also provide for reception, demodulation, and parameter measurements of the Centaur S-band telemetry signal in the Cx 36 GTS.

SECTION 7

OPERATIONS

Development of a preliminary operations plan for Centaur G-prime has been coordinated with KSC ground operations personnel and JSC flight operations personnel. This plan is compatible with planned Shuttle ground and flight operations. Safety is a prime consideration in all operations.

A Preliminary Flight Operations Plan, Payload Integration Plan, and Launch Site Support Plan have been presented to JSC and KSC.

System Interface Requirements (SIR) documents have been prepared defining interface requirements at the Vertical Processing Facility (VPF), Orbiter Processing Facility (OPF), Launch Complex 39A/B, and the multi-use mission support equipment canister (MMSE), General Dynamics personnel at the Eastern Space and Missile Center (ESMC) will provide continuing liaison with KSC.

A review copy of the Support Instrumentation Requirements Documents (SIRD) for the Centaur upper stage has been prepared defining baseline support requirements from the Tracking and Data Relay Satellite System (TDRSS) and the NASA Communications Network (NASCOM) during the Centaur free-flight phase.

Preliminary flight operations event, time-lines, and crew functions have been defined. This data is compatible with Orbiter operations. The necessary analyses will be performed to develop the required documentation and procedures.

7.1 GROUND OPERATIONS

General Dynamics has extensive experience in working with NASA in ground operations involving Centaur vehicles at Complex 36A and 36B (Centaur D-1A) and Complex 41 (Centaur D-1T).

7.1.1 NORMAL OPERATIONS. Before receiving flight hardware at the Eastern Launch Site, the avionics simulators will be used to verify compatibility with the support equipment at Complex 36, VPF, and Complex 39. Following propellant skid validations at Complex 36A, the skids will be installed at Complex 39A/B and cryogenic cold flow tests performed to validate propellant loading interfaces.

Figures 7-1 through 7-3 show the flow of Centaur and its support equipment from receipt at Cape Canaveral Air Force Station through launch at KSC. Two CLLSs, two MSE trailers, two FSE sets (39A & 39B), three LO_2 control skids, two LH_2 control skids, and three helium control skids will be used to support the Galileo and ISPM operations.

Above: A montage of Concept art showing deployment of a Centaur G-Prime and its interplanetary probe from the Space Shuttle. Black and white version of color image No. C-1983-187, courtesy NASA/Lewis Research Center.

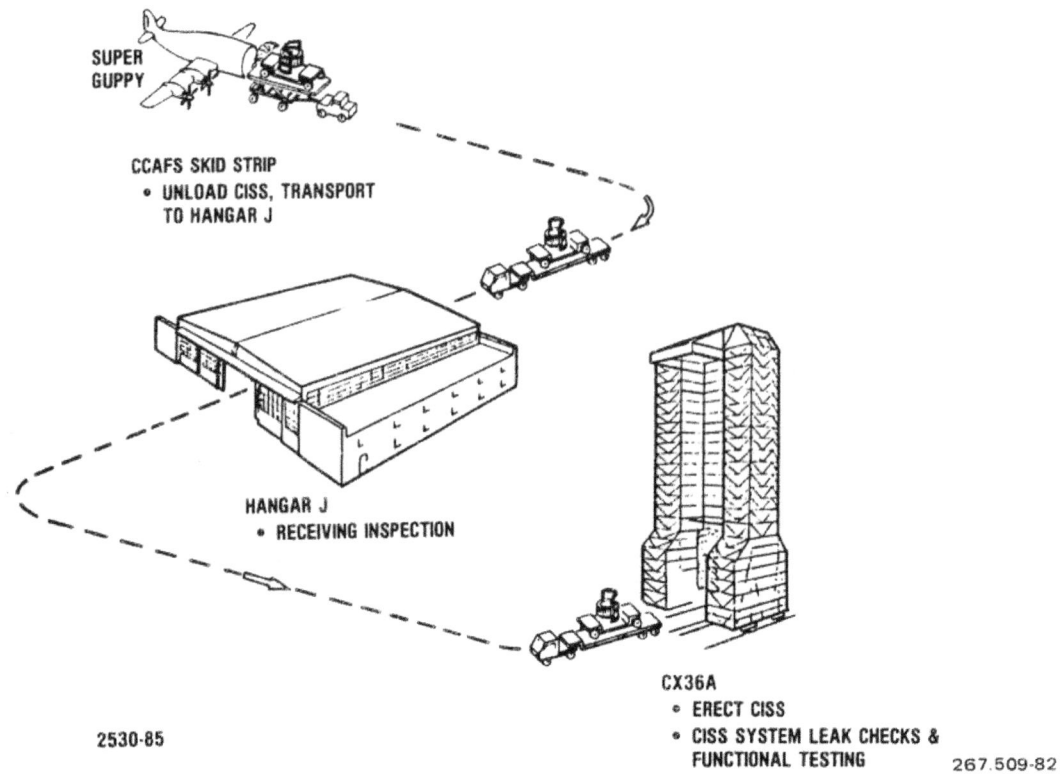

Figure 7-1. Centaur G-Prime CISS Prelaunch Flowpath

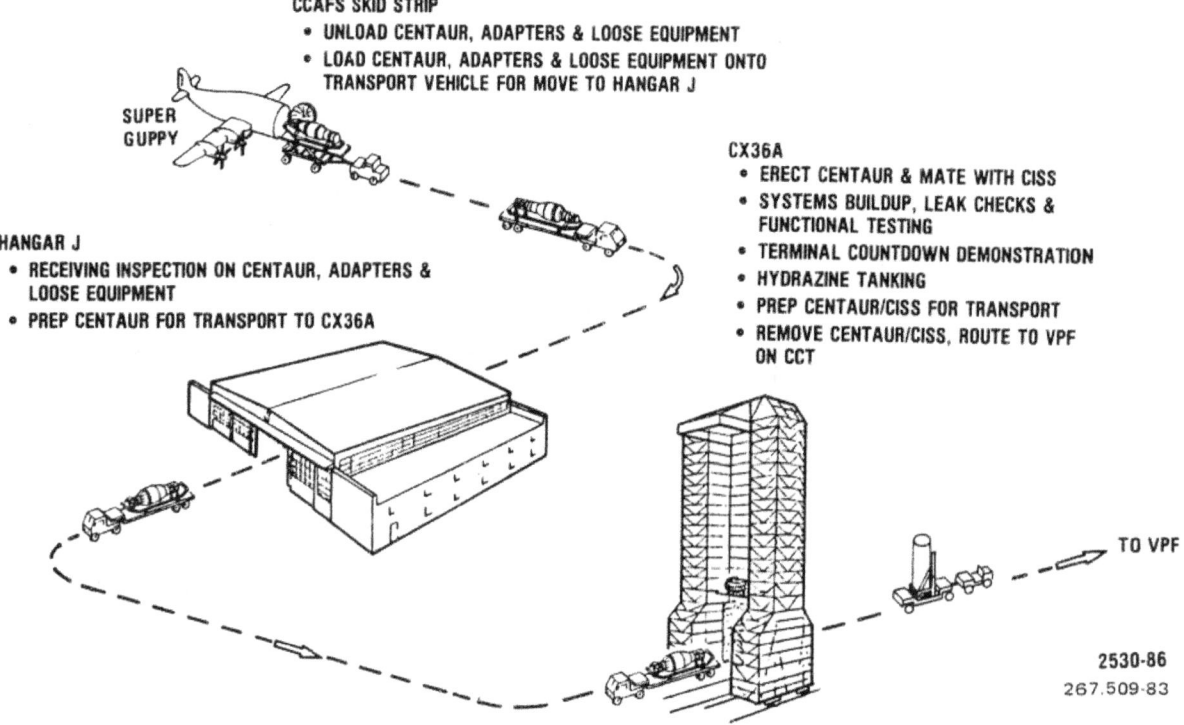

Figure 7-2. Centaur G-Prime Processing Flow at CCAFS Facilities

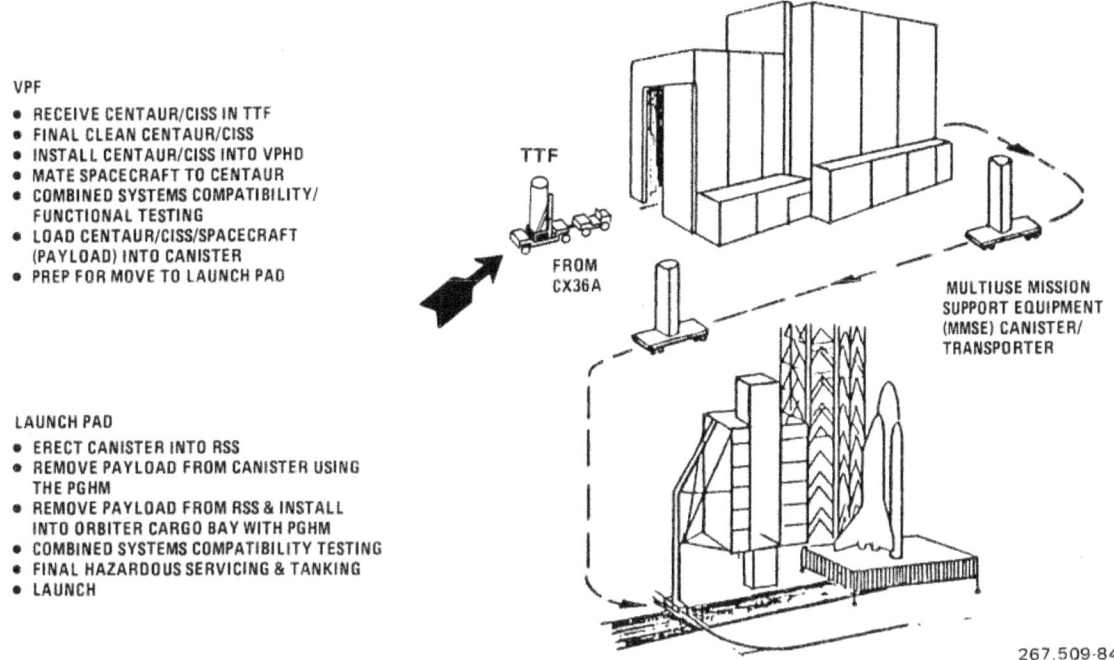

VPF
- RECEIVE CENTAUR/CISS IN TTF
- FINAL CLEAN CENTAUR/CISS
- INSTALL CENTAUR/CISS INTO VPHD
- MATE SPACECRAFT TO CENTAUR
- COMBINED SYSTEMS COMPATIBILITY/
  FUNCTIONAL TESTING
- LOAD CENTAUR/CISS/SPACECRAFT
  (PAYLOAD) INTO CANISTER
- PREP FOR MOVE TO LAUNCH PAD

LAUNCH PAD
- ERECT CANISTER INTO RSS
- REMOVE PAYLOAD FROM CANISTER USING
  THE PGHM
- REMOVE PAYLOAD FROM RSS & INSTALL
  INTO ORBITER CARGO BAY WITH PGHM
- COMBINED SYSTEMS COMPATIBILITY TESTING
- FINAL HAZARDOUS SERVICING & TANKING
- LAUNCH

TTF

FROM
CX36A

MULTIUSE MISSION
SUPPORT EQUIPMENT
(MMSE) CANISTER/
TRANSPORTER

267.509-84

Figure 7-3. Centaur G-Prime Flow Path at KSC

Following Complex 36A testing, which includes a cryotanking, the Centaur/CISS is transported to the VPF and mated with the spacecraft. Orbiter interface validation using the cargo integration test equipment (CITE) is performed. When these tests are completed, the spacecraft/Centaur/CISS (cargo) is installed in the MMSE canister and transported to the Complex 39 Rotating Service Structure (RSS). The cargo is subsequently installed into the Orbiter bay using the RSS payload ground handling mechanism. A Countdown Demonstration Test will be performed to validate the cargo Shuttle/Complex 39A/B combination. Final integration testing and prelaunch operations are performed, and the Shuttle is launched.

After Orbiter landing, the CISS will be removed from the Orbiter while in the OPF and returned to Hangar J as shown in Figure 7-4.

7.1.2 ABORT OPERATIONS. The Centaur vehicle is an integral part of the Orbiter safing and securing procedure following the return from an abort (Figure 7-5). At the Shuttle Landing Facility, the Centaur/CISS system status will be assessed via telemetry data. Within approximately 30 minutes of Orbiter landing, access will be required to the Orbiter aft fuel T-0 panel to connect a gaseous helium charge line to replenish the CISS helium supply. Following confirmation that all safety requirements are met, the Orbiter is towed to the OPF.

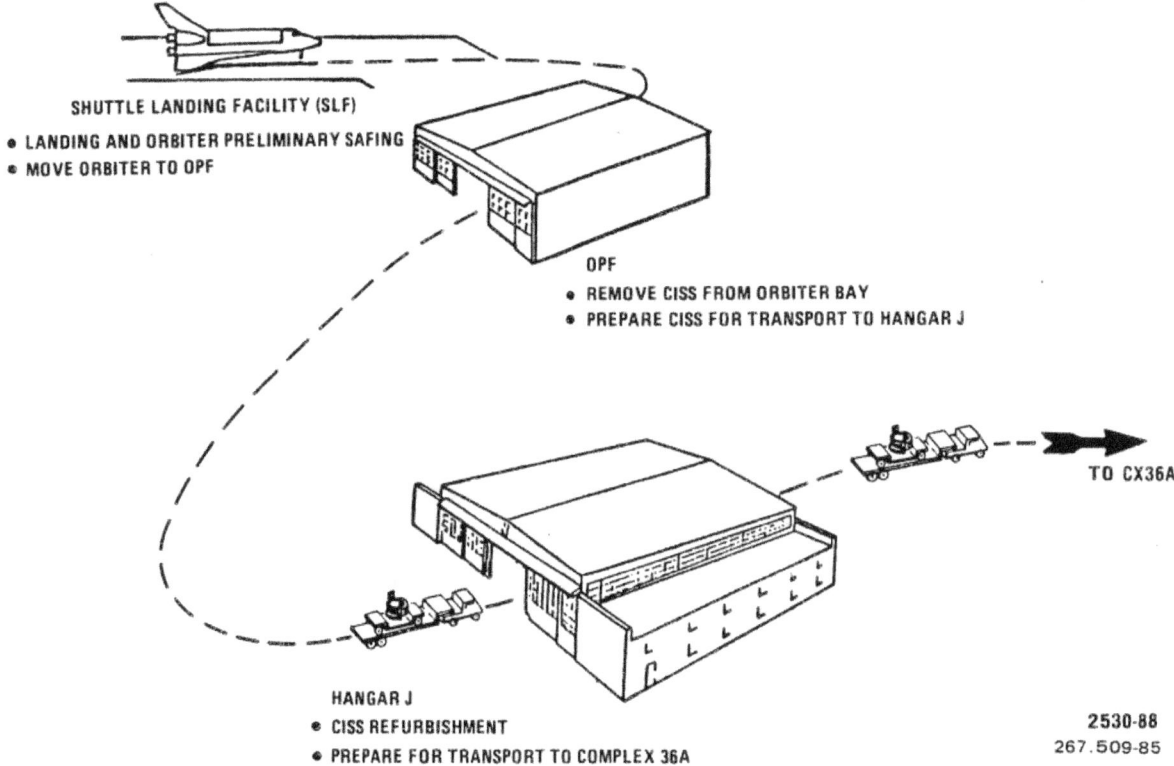

Figure 7-4. Post Flight CISS Flowpath

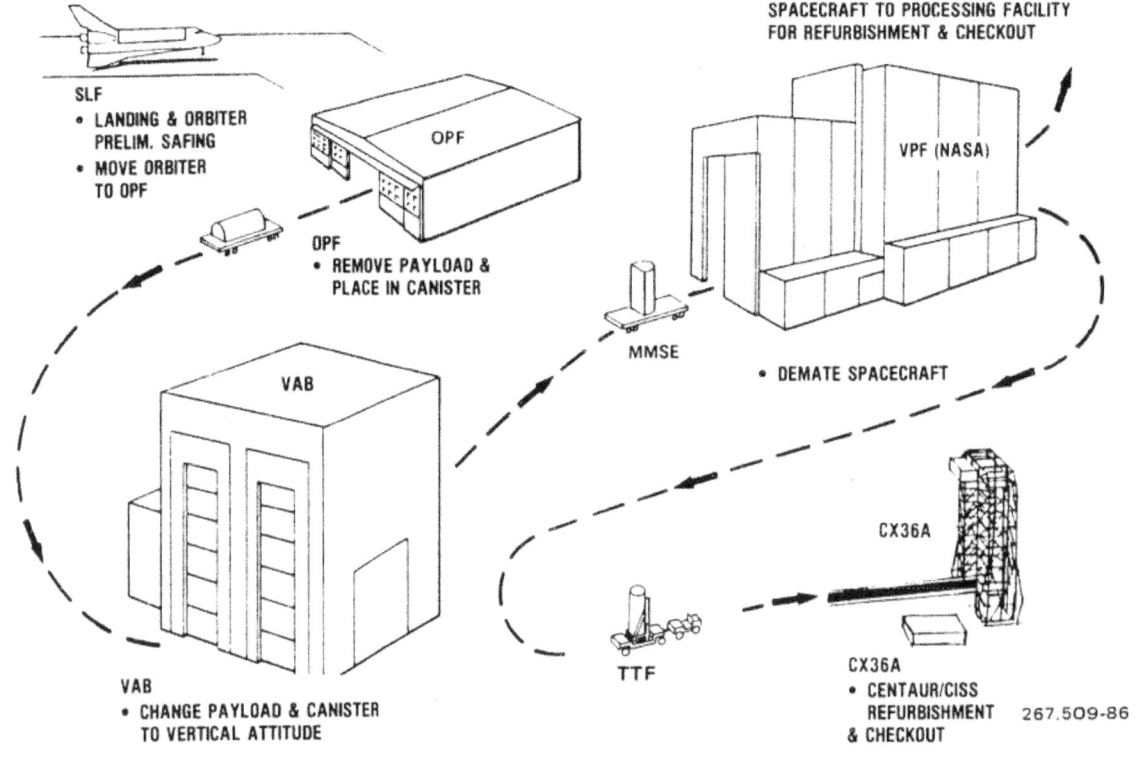

Figure 7-5. Centaur G-Prime Abort Flow Path

In the OPF, the Orbiter is prepared for Centaur/CISS/spacecraft assembly removal. The assembly is removed and placed horizontally in the MMSE canister. It is then moved to the Vehicle Assembly Building where the canister is rotated to the vertical position. In this configuration the assembly can be returned to the VPF for spacecraft removal and subsequent refurbishment and checkout as required.

7.2 FLIGHT OPERATIONS

Centaur G-prime will perform the NASA interplanetary missions using flight operations plans and procedures that mesh smoothly into those for Mission Control and the Orbiter crew.

Centaur G-prime will fly the Galileo and ISPM missions within the NASA and JPL flight operations requirements. Flight operations planning and support for other missions will be similar. These operations will be developed as required to support individual program requirements.

The generic and mission-peculiar flight operations requirements include:

- Centaur G-prime will accomplish the reference trajectory with a figure of merit (FOM) equal to or better than the 24 msec requirements for the Galileo spacecraft (35 msec for ISPM) over a 10-day launch window. This will be done while operating within the constraints of the flight operations requirements.

- Preflight planning will coordinate training, operations control, operations support, and flight plans for nominal, contingency, and abort missions.

- Flight operations - nominal, contingency, and abort - will insure coordinated activities of the CCE, Orbiter, and crew with one another and with operations support on the ground.

- Flight operations during flight will support the command and control for Shuttle/Centaur. This includes ground-based activities ranging from data monitoring and operations advice to the MCC, to operational control during Centaur free flight.

- Shuttle abort-to-orbit (ATO) may include near-nominal CCE operations, since the Centaur G-prime can deploy and perform a successful Galileo mission from the 105-n.mi. ATO altitude.

- Postflight evaluation will verify data and operations related to the CCE, and recommend corrective action as necessary.

7.2.1 PREFLIGHT PLANNING. Centaur cargo element (CCE) flight operations plans will be developed for each mission. (The CCE includes the CISS, Centaur G-prime vehicle, and the spacecraft.) The overall objectives of the Centaur G-prime flight operations are to:

1. Make the Centaur cargo flight operation-peculiar items as independent of the Orbiter operations as possible

2. Give the Orbiter crew control initialization of such critical functions as deployment, arm/safe, and abort operations

3. Coordinate and integrate Centaur G-prime operations with the spacecraft, Orbiter flight crew, and Mission Control Center operations.

Centaur Flight Operations will be integrated into the total Shuttle Flight Operations Plan.

The basic purpose of Centaur Flight Operations is to integrate operations of the various CCE subsystems to meet the mission requirements and to be compatible with Shuttle operational requirements. Elements of the plan include ground rules and assumptions, a sequence events/mission timeline, procedures, and command, control, and communication links. The plan also includes detailed abort section with operational backout procedures.

The Flight Operations activity will be used to develop input data for the basic STS Payload Integration Plan (PIP) and several PIP Annexes: Flight Planning Annex, Training Annex, Payload Operations Control Center (POCC) Annex, Orbiter Crew Compartment Annex, Command & Data Annex, and the Flight Operations Support Annex. The latter requires development of flight rules, decision points, and alternative plans and guidelines for off-nominal flight operations.

7.2.2 <u>FLIGHT OPERATIONS SUPPORT</u>. During ascent and Orbiter-attached operations, the Galileo and ISPM and flight operations will be controlled from the MCC. During this period, the Multi-Purpose Support Room (MPSR) will support CCE operations for the MCC at JSC, as shown in Figure 7-6. The Centaur Payload Operations Control Center (CPOCC) will support the MCC and be in close communication with the MPSR. CPOCC will also have prime responsibility for monitoring Centaur systems

Within the MCC, the Flight Control Room (FCR) will control mission operations. Centaur flight controllers in the MPSR will include GDC Centaur specialists to support the activities of the FCR during Orbiter-attached operations. After separation, when the Orbiter has maneuvered out of the zone of safety around the Centaur, control of the Centaur will be handed over to the CPOCC. The CPOCC will then control Centaur flight operations through Centaur burn, spacecraft separation, and Centaur post-separation maneuvers.

During the ascent phase of flight, flight operations support consists primarily of monitoring CCE health status until the Orbiter payload bay doors are opened. On-orbit operations then being with initiation of Centaur/CISS checkout and spacecraft checkouts (if required).

Flight operations support continues on-orbit with ground analysis of checkout data, continuous real-time analysis of CCE health status, and providing advice regarding Centaur as necessary to the FCR. The latter includes go/no-go decisions for Centaur rotation and separation from the Orbiter.

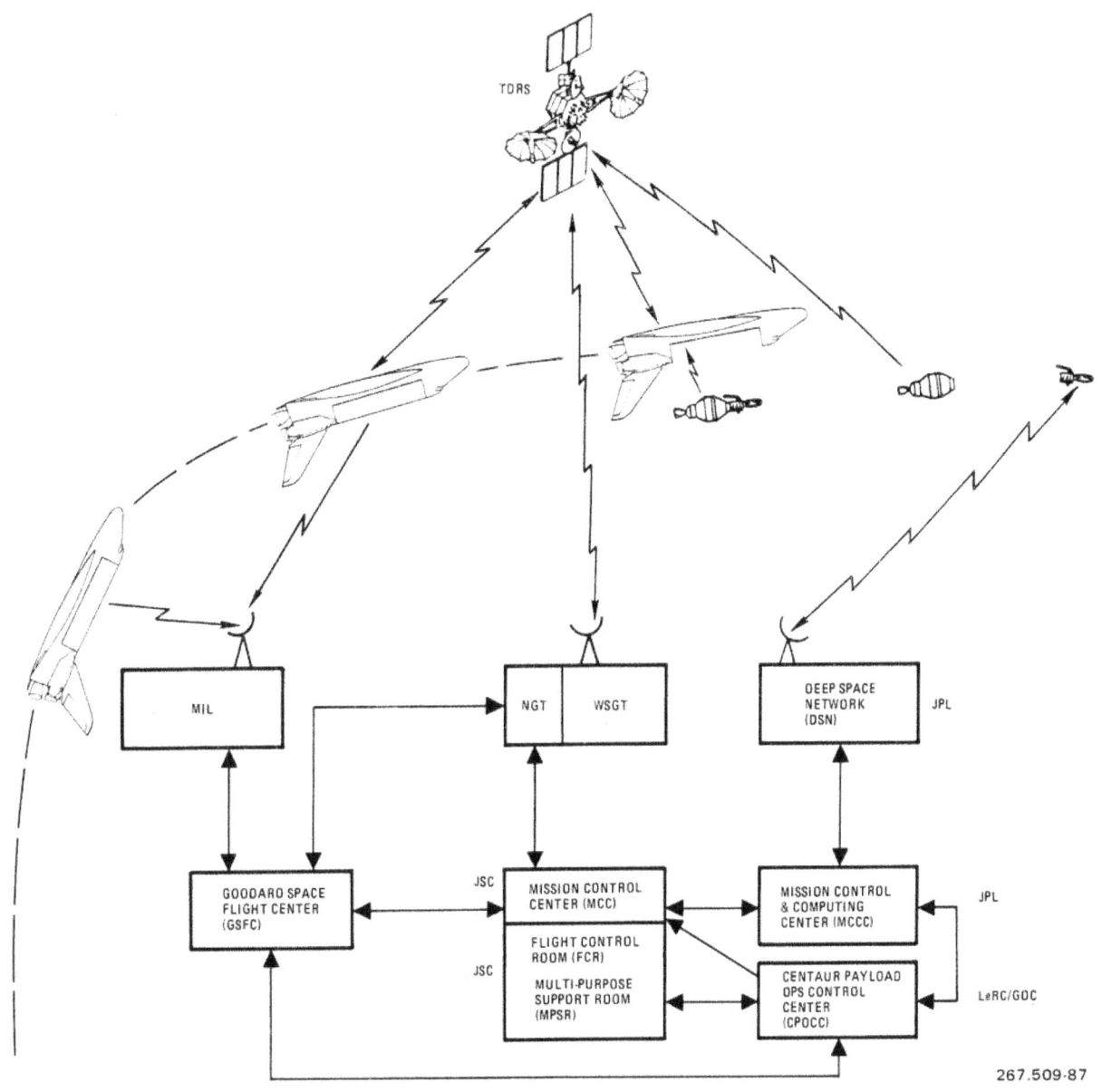

Figure 7-6. CPOCC Support for NASA Mission Flight Operations

Flight operations support of the Centaur vehicle after handover to the CPOCC
consists primarily of monitoring Centaur automatic sequences, data recording
and evaluating anomalies.

Flight operations requirements include CPOCC staffing, telemetry links,
decision points, and ground analysis of data.

7.2.3 <u>NOMINAL FLIGHT</u>. Nominal CCE flight operations are continuous from launch through orbit injection and post-separation maneuvers. The CCE is limited to safety functions such as passive navigation, vent control, and pressurization control until the crew assumes an active role in CCE on-orbit pre-deployment operations. Based on Shuttle flight requirements for ascent phase and on-orbit reconfiguration, the earliest time that CCE on-orbit pre-deployment operations can begin is 1 hour 13 minutes (all times referenced to liftoff). Nominal flight operations from this point on are illustrated in Figures 7-7 through 7-10.

From liftoff through Centaur post-deployment operations, the MCC and CPOCC will monitor the health and safety of the CCE via the telemetry link. The crew also has independent access to status information via a CRT display and can act as a backup source during attached operations.

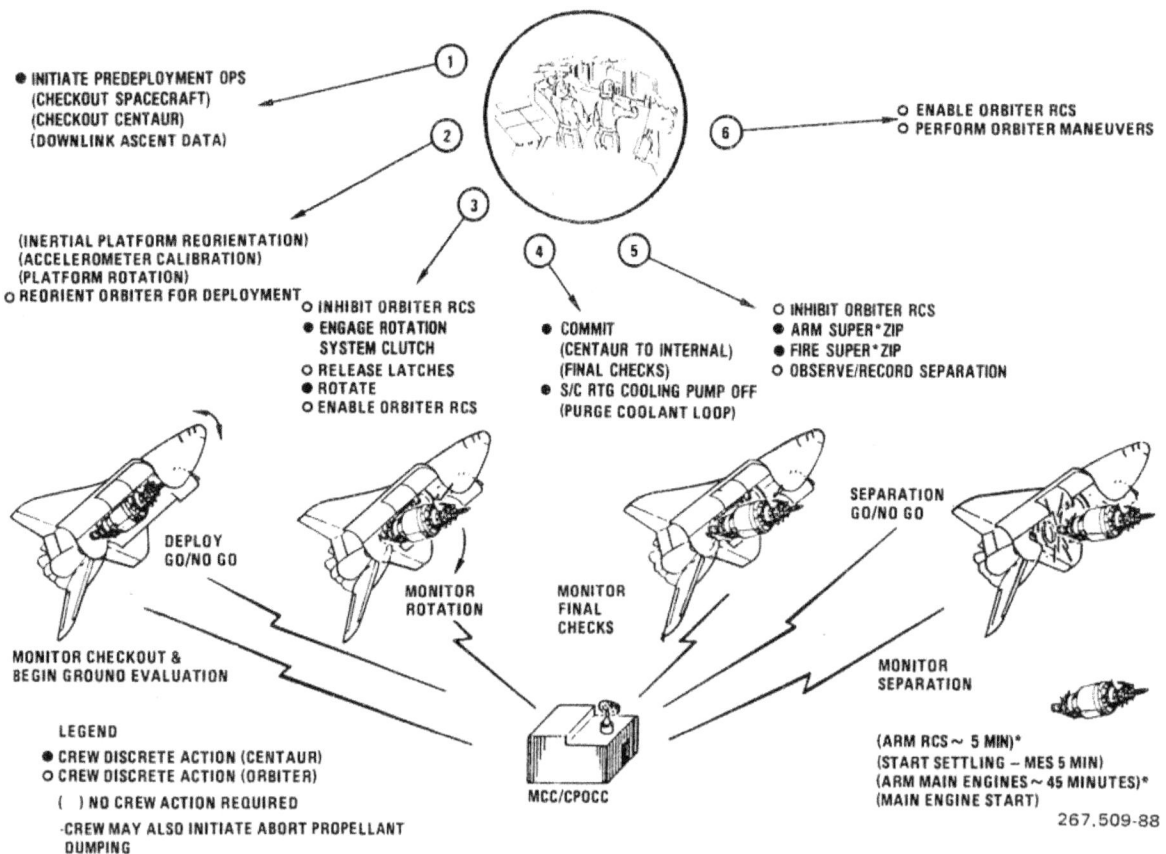

Figure 7-7. Orbiter Crew Control of Critical Functions

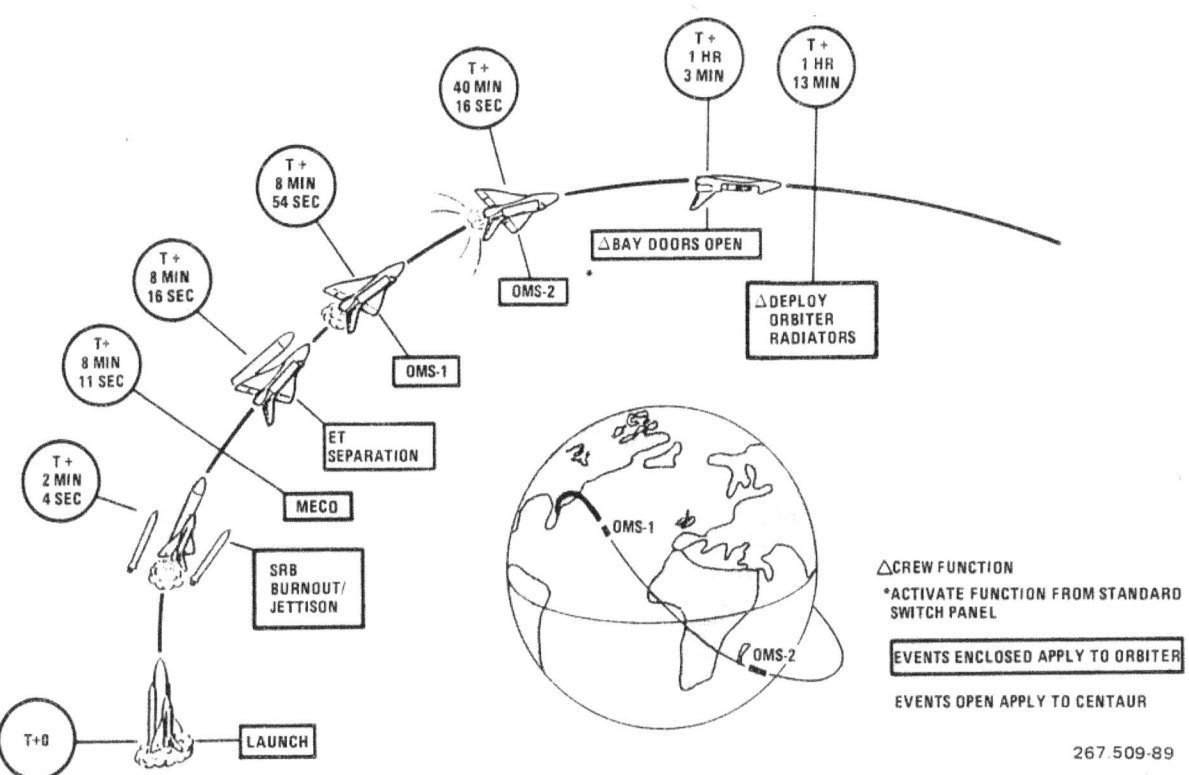

Figure 7-8. Nominal CCE Ascent Operations

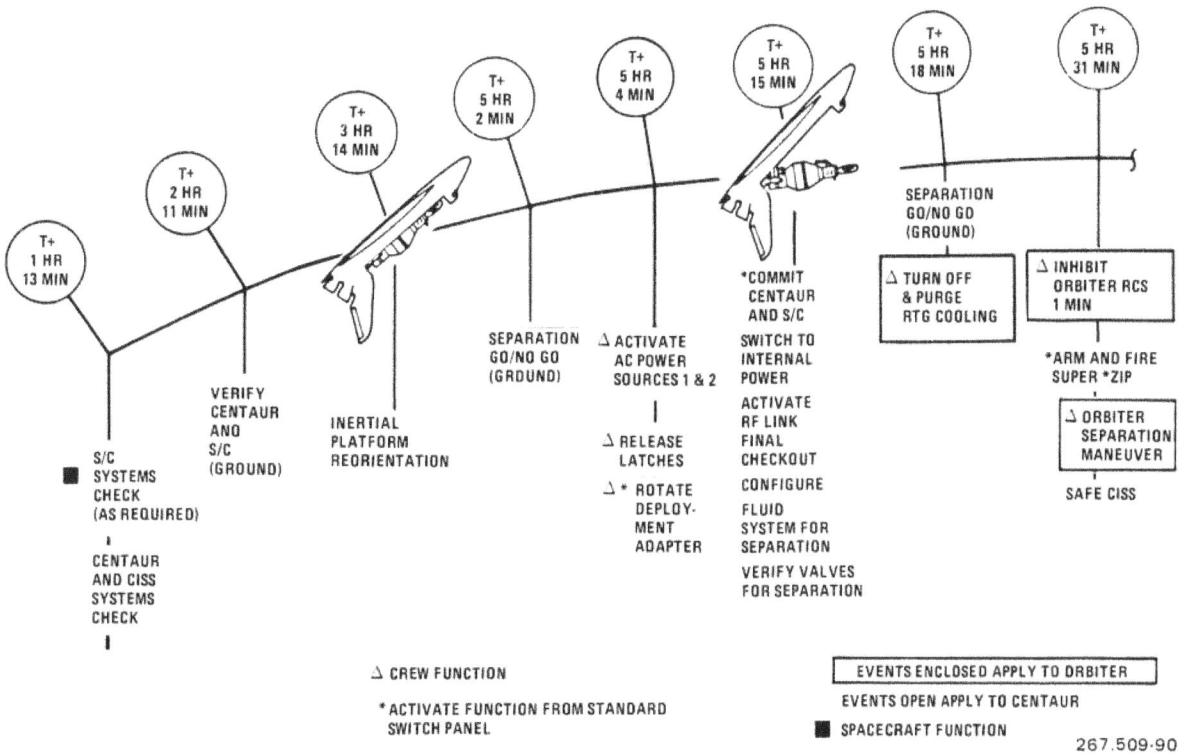

Figure 7-9. Rev 4 Deployment Timeline

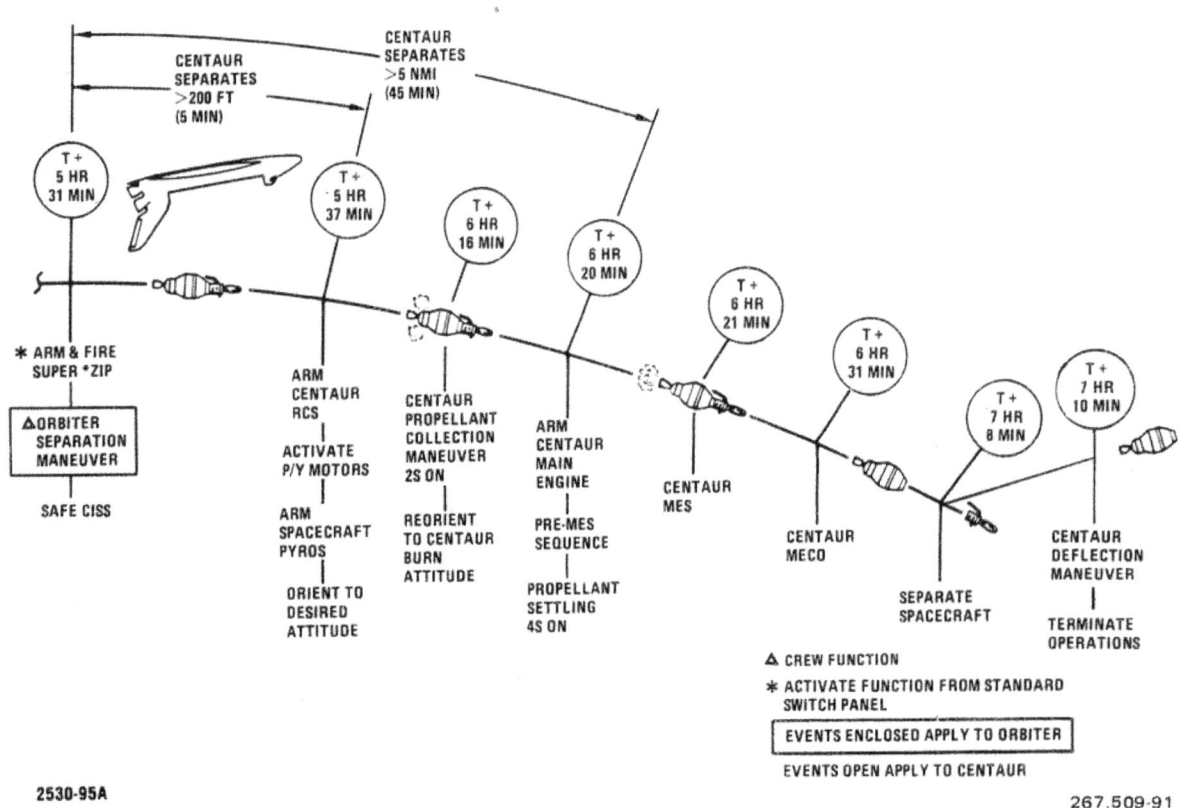

CENTAUR SEPARATES >200 FT (5 MIN)

CENTAUR SEPARATES >5 NMI (45 MIN)

T + 5 HR 31 MIN

T + 5 HR 37 MIN

T + 6 HR 16 MIN

T + 6 HR 20 MIN

T + 6 HR 21 MIN

T + 6 HR 31 MIN

T + 7 HR 8 MIN

T + 7 HR 10 MIN

＊ ARM & FIRE SUPER ＊ZIP

△ORBITER SEPARATION MANEUVER

SAFE CISS

ARM CENTAUR RCS

ACTIVATE P/Y MOTORS

ARM SPACECRAFT PYROS

ORIENT TO DESIRED ATTITUDE

CENTAUR PROPELLANT COLLECTION MANEUVER 2S ON

REORIENT TO CENTAUR BURN ATTITUDE

ARM CENTAUR MAIN ENGINE

PRE-MES SEQUENCE

PROPELLANT SETTLING 4S ON

CENTAUR MES

CENTAUR MECO

SEPARATE SPACECRAFT

CENTAUR DEFLECTION MANEUVER

TERMINATE OPERATIONS

△ CREW FUNCTION

＊ ACTIVATE FUNCTION FROM STANDARD SWITCH PANEL

EVENTS ENCLOSED APPLY TO ORBITER

EVENTS OPEN APPLY TO CENTAUR

2530-95A

267.509-91

Figure 7-10. Nominal/Centaur Postdeployment Operations

Figure 7-7 illustrates the Orbiter crew control/interface for the flight operations by means of the standard switch panel or through the Orbiter keyboard. Switch-initiated actions are identified by asterisks in Figures 7-8 and 7-9, along with an outline of the computer-controlled automatic sequences.

Figure 7-9 shows the Centaur checkout and deployment operations required to deploy a Centaur with the Galileo spacecraft. Also shown are the Orbiter support operations required (e.g., reorientation maneuver, Orbiter RCS inhibit times, etc.), Orbiter crew control functions (denoted by asterisks), and two go/no-go key decision points: (1) to initiate the rotation operation at about 30 minutes before Centaur separation, and (2) at the commit sequence just prior to separation.

The deployment timeline is based on the reference Galileo mission, with the Centaur main engine start (MES) consistent with mission requirements.

All flight operations requirements, both generic and mission-peculiar, as listed in Table 7-1, are met by this sequence. This includes requirements for crew initiation of events, Centaur platform rotation and accelerometer bias calibration to meet the mission FOM, deployment attitude thermal control for the spacecraft, and Centaur separation in daylight.

Table 7-1. Centaur G-Prime Flight Operations Requirements

Flight Operations - Generic

1. Orbiter (099 or 104) will be the Shuttle Orbiter modified for Centaur. GLL takes advantage of the additional performance capability of 104.

2. Centaur/spacecraft shall be a dedicated payload, and as such, need not consider other payload/requirements during the same STS mission.

3. Centaur flight operations shall provide for crew initiation of critical events. The events include Centaur unlatch, rotation (up and down), and separation.

4. The crew will switch the CCE power to the Orbiter power supply after ascent if the full 3.2 kW is not available from the Orbiter during the ascent phase.

5. Centaur will perform an inertial platform rotation while attached to the Orbiter that will occur midway between launch and Centaur MECO.

6. Orbiter OMS and translation RCS shall be inhibited for deployment adapter rotation, and for Centaur separation. Orbiter primary and vernier RCS is to be inhibited during Centaur separation.

7. Centaur separation shall occur in daylight.

8. Centaur separation system shall impart a minimum of 1 fps relative velocity between the Centaur and Orbiter. (The RMS will not be used.)

9. Centaur RCS shall not be armed until a distance of 200 ft (an elapsed time of 5 minutes) has been achieved after separation from Orbiter.

10. Centaur propellant tank pressurization and vent systems shall be inhibited before separation until after Centaur RCS is enabled.

11. No provisions shall be made for the Orbiter crew and ground operations to inhibit and re-enable the vehicle arming sequence via RF link.

12. Centaur main engine system shall not be armed until an elapsed time of 45 minutes has been achieved after separation from the Orbiter.

Flight Operations - Mission Peculiar

13. The ISPM and Galileo missions shall be launched in May - June 1986.

14. Spacecraft attitude-thermal constraints are: (1) no direct sunlight in Orbiter payload bay with short duration exceptions, and 2) a thermal slow-roll (0.1 rpm) until RTG booms are deployed.

15. During postseparation, pre-MES coast, (TBD) spacecraft discretes (Galileo only) are to be issued by Centaur.

Table 7-1. Centaur G-Prime Flight Operations Requirements, Contd

16. Centaur shall reorient for separation then achieve a high spin (2.9 rpm; dependent upon S/C mass properties/cross products of inertia).

Contingency Operations - Generic

17. In event of an abort, Centaur main propellants (LO_2 and LH_2) shall be dumped before return.

18. Orbiter RCS propellant settling thrust shall be provided to initiate and terminate Centaur propellant dump for AFO abort. The Orbiter OMS thrust shall be used for Centaur propellant dump for AOA aborts.

19. Centaur systems shall permit simultaneous dump of LO_2 and LH_2 propellants within 250 sec during RTLS and TAL abort.

20. Centaur propellant abort dump shall be automatic during RTLS, TAL and AOA aborts, with a manually initiated backup option.

21. Centaur is to continue mission in case of an ATO abort that does not affect capability to complete mission successfully.

22. Abort preferences are: ATO over AOA, and AOA over TAL.

23. Centaur systems shall accommodate a worst-case, closed-door abort thermal environment. Such an abort consists of the Orbiter payload bay doors remaining closed continuously with the duration from liftoff to landing of 6.5 hours.

24. Mission will be aborted if cargo bay doors are not opened within three hours MET.

Contingency Operations - Mission Peculiar

25. Transmit signal from Centaur to spacecraft in case of on-orbit abort. RTG cooling shall not be terminated.

Flight Support Operations - Generic

26. A CPOCC provided by NASA and staffed by General Dynamics and NASA-LeRC shall be used to monitor Centaur operations and provide inputs to the mission director on Centaur health status throughout attached and detached phases of flight.

27. TDRSS will be available and used for S-band and phase-modulated (PM) and Ku-band downlink.

28. GSTDN ML station will be used for S-band downlink on ascent.

Table 7-1. Centaur G-Prime Flight Operations Requirements, Contd

29. Centaur/CISS data must be downlinked in real time to MCC/CPOCC. Quantity of data downlinked to be maximized within communications limitations and operational requirements of Orbiter and communications system.

30. Ascent data will be recorded on the payload data recorder and downlinked as soon as communications constraints allow.

31. A period of 60 minutes (minimum) will be allowed after checkout of the Centaur and CISS before a go/no-go decision for rotation is transmitted from the CPOCC to the MCC for relay to the crew.

32. A maximum period of three minutes will be allowed after rotation and final checkout before go/no-go decision for separation is transmitted from the CPOCC to the MCC for relay to the crew.

33. Centaur data link will be transferred from the Orbiter to the TDRSS as soon as possible after separation of the Centaur from the Orbiter.

34. Centaur flight control operations will be transferred to the CPOCC as soon as Centaur transmits direct to TDRSS.

Flight Support Operations - Mission Peculiar

35. Orbiter Ku-band will not be used when incident RF on the spacecraft is greater than 47 volts per meter.

36. Spacecraft on-orbit checkout of 120 minutes will be required for ISPM and may be halted and restarted in segments by command (segments to be 20 min or greater).

37. Spacecraft real-time telemetry will be required during Centaur deployment (30 minutes), burn, spinup through separation, and separation through DSN acquisition.

Figure 7-10 illustrates operations from Orbiter/Centaur separation to Galileo spacecraft separation. Centaur coast attitude and sequence times are appropriate to meet spacecraft thermal constraints. Centaur separation coast and inhibit/arm sequence times are sufficient to ensure Orbiter safety constraints before MES. If the deployment opportunity is lost, additional deployment opportunities with Centaur may be realized.

Three Orbiter post-deployment operations are required after Centaur separation: (1) Orbiter maneuvers away from the Centaur without contaminating the spacecraft, (2) Orbiter monitors Centaur downlink until RF transfer direct to TDRSS, and (3) the CISS will be safed for atmospheric reentry and landing (e.g., venting the pressurant bottles and pressurizing lines to atmospheric levels).

7.2.4 ABORT. The primary abort objective is to land safely with the space-craft, Centaur, and CISS intact and reusable (after refurbishment) for a later flight. The Centaur/spacecraft can be restowed in the Orbiter payload bay up to the time of physical separation by the Super*Zip separation ring.

CCE flight operations will be developed for five pre-planned Shuttle abort modes:

1. Return to Launch Site (RTLS).

2. Trans-Atlantic Abort Landing (TAL).

3. Abort Once Around (AOA).

4. Abort to Orbit (ATO).

5. Abort From Orbit (AFO).

A deployment backout sequence will be developed that can be performed at any time from a normal deployment sequence. Once initiated by the crew, this backout sequence will be computer controlled, leaving the Orbiter crew free to attend to Orbiter operations; however, a certain degree of CCE support may be required to manually backup abort operations, e.g., initiate Centaur propellant dump.

To return with the CCE intact, a Centaur propellant dump is planned before re-entry for AOA and AFO aborts, or before Orbiter main engine cutoff for RTLS and TAL aborts. Analyses of this propellant dump capability have been made for all preplanned Shuttle abort modes, as illustrated in Figure 7-11.

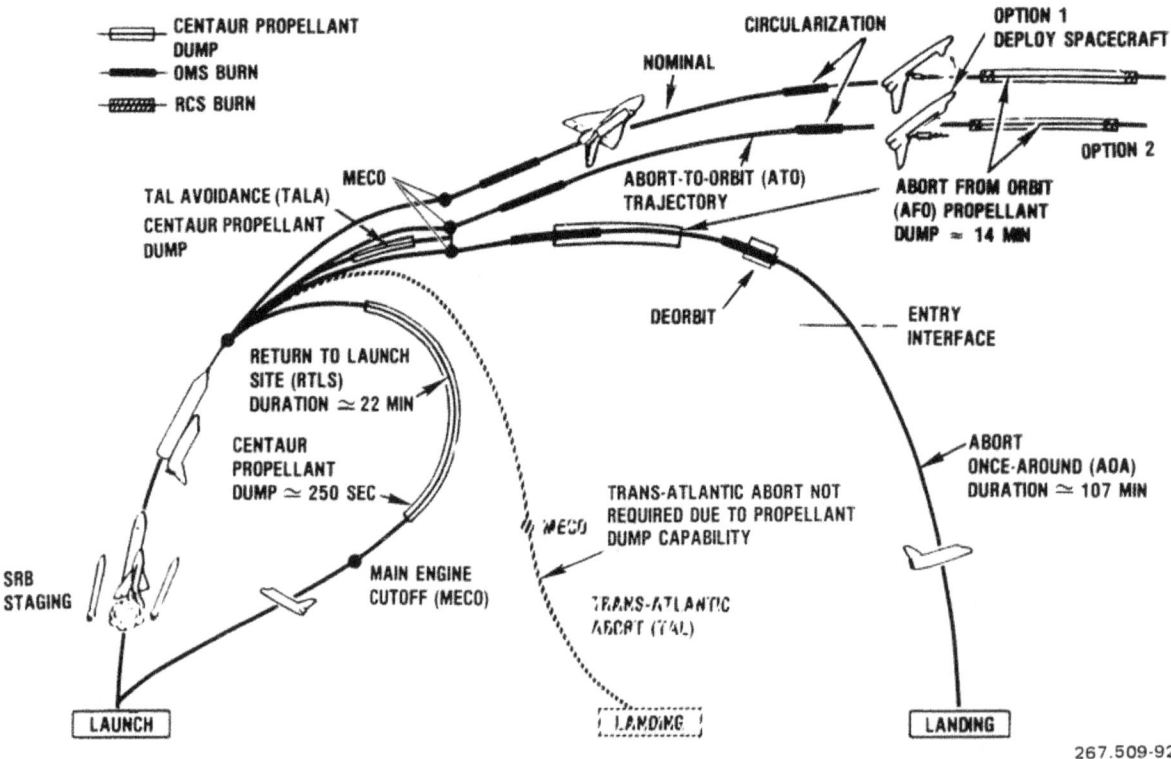

267.509-92

Figure 7-11. Centaur Can Dump Propellants Safely in All Abort Modes

Centaur nominal dump time of 250 seconds during RTLS and TAL aborts is designed to be completed before MECO. (The TAL mode may be virtually eliminated by dumping Centaur propellants and continuing into an ATO or AOA mode.) The dump system can operate in a zero-g environment (on-orbit, Figure 7-12), but requires a settling acceleration at the start and end of the dump to minimize residual propellants. These settling burns may be provided by either the Orbiter RCS (AFO mode) or by the orbital maneuvering system (OMS) during normal OMS burns (AOA mode).

An abort to orbit may still result in a successful mission. When an ATO is caused by non-mission-critical functions, the Orbiter can remain in its 105-n.mi. ATO orbit from five orbital revolutions up to one day; consequently, Centaur can proceed with deployment and perform the mission.

All planned abort operations are in accordance with contingency requirements of Table 7-1.

The STS caution and warning (C&W) system will require CCE inputs. The criteria for issuing a C&W signal will be programmed into the safety and health status software of the CUs. This evaluation will be provided to the Orbiter crew and to the ground MCC/CPOCC; specific systems will be identified and their tolerance conditions indicated. The crew and ground support can also review the status data on any CRT display to highlight critical items. Alternative flight plans will be developed for such possible contingency conditions and will be included in the Flight Operations Support Annex.

7.2.5 POSTFLIGHT EVALUATION. Each flight will have a detailed evaluation to verify correct operation of all Centaur/CISS systems. Any anomaly will be evaluated and corrective action taken before the next flight.

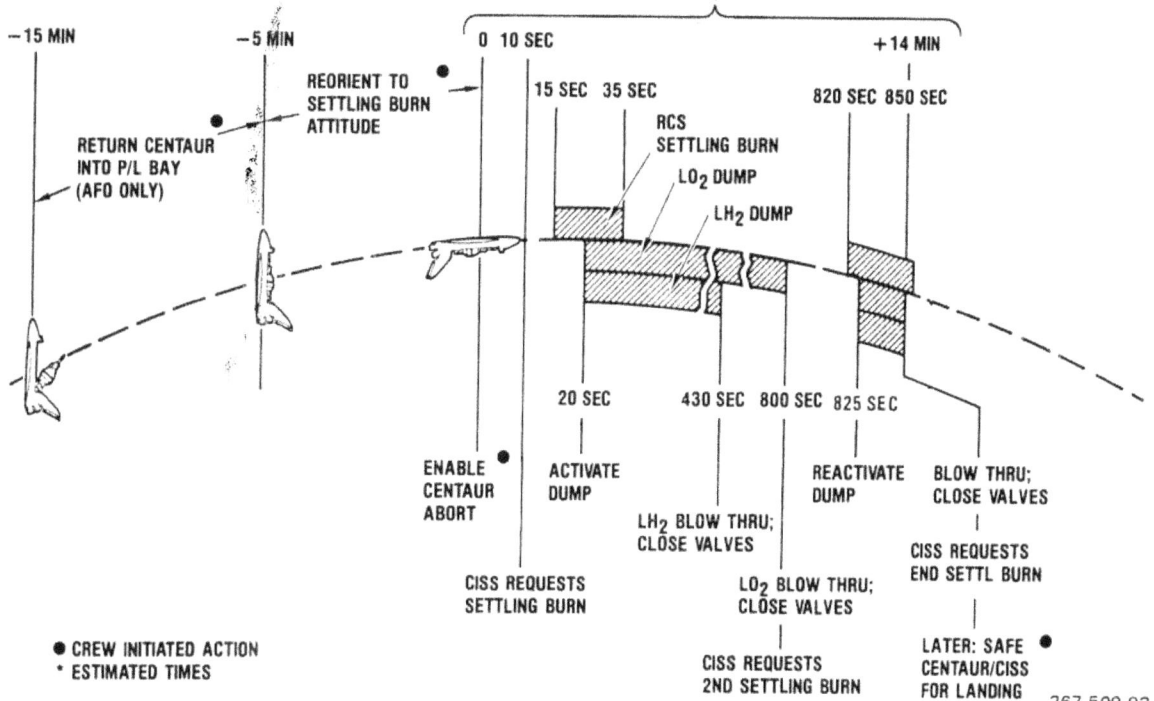

Figure 7-12. Zero-G Propellant Dump

Above: Concept art showing deployment of a Centaur G-Prime and its interplanetary probe from the Space Shuttle. Black and white version of color image No. C-1983-187, courtesy NASA/Lewis Research Center.

SECTION 8

DEVELOPMENT REQUIREMENTS

8.1 MAJOR ANALYSIS REQUIREMENTS

Accuracy of this analyses is demonstrated by GDC's 38 consecutive operational successes in Centaur flights since 1971.

8.1.1 <u>STRUCTURAL</u>. Centaur G-prime structural sizes and interface loads have been determined and verified to be acceptable through the use of detailed finite-element models and dynamic transient analyses based on Space Transportation System (STS) models and the forcing functions.

Orbiter interface locations chosen were those with the highest strength between frames in the Orbiter bay, based on ICD-2-1F001. Interface loads and point accelerations were initially calculated using the preliminary design accelerations of ICD-2-1F001. These interface loads and accelerations were then used, along with detailed finite element models (Figures 8-1 and 8-2), discontinuity analyses, buckling analyses, or standard frame solutions of each major piece of structure to determine optimized section sizes of the tank, adapters, and CISS.

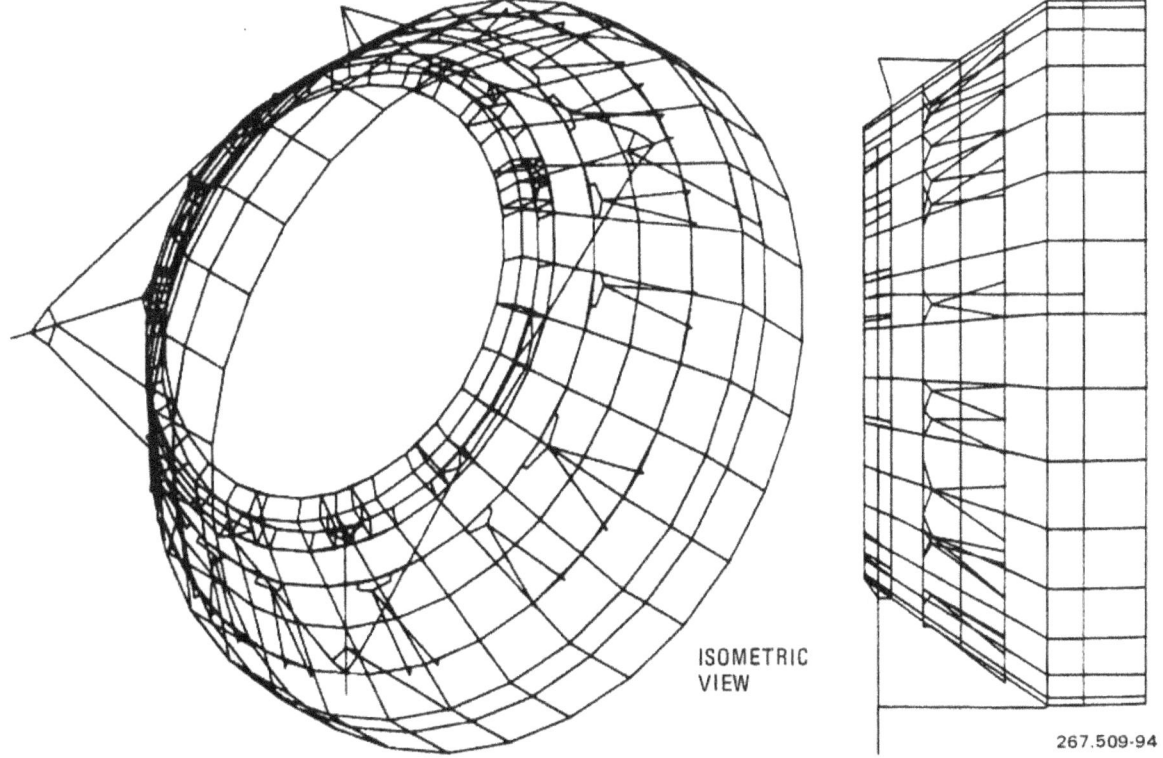

ISOMETRIC
VIEW

267.509-94

Figure 8-1. Centaur G-Prime Forward Adapter Finite Element Model

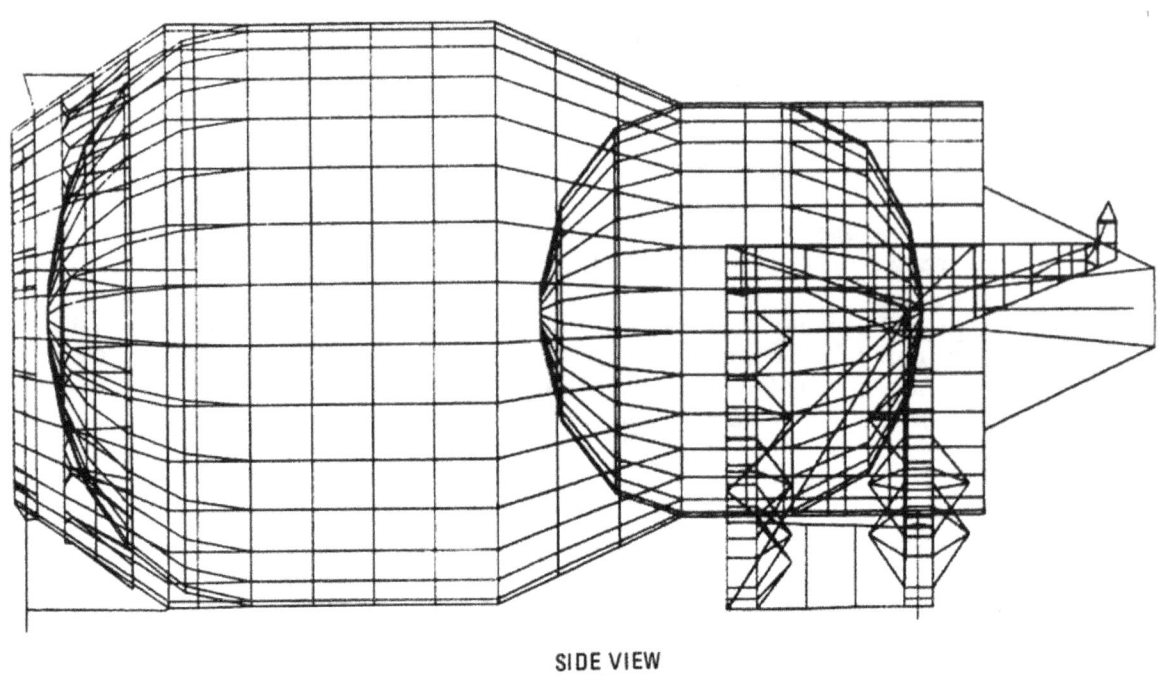

SIDE VIEW

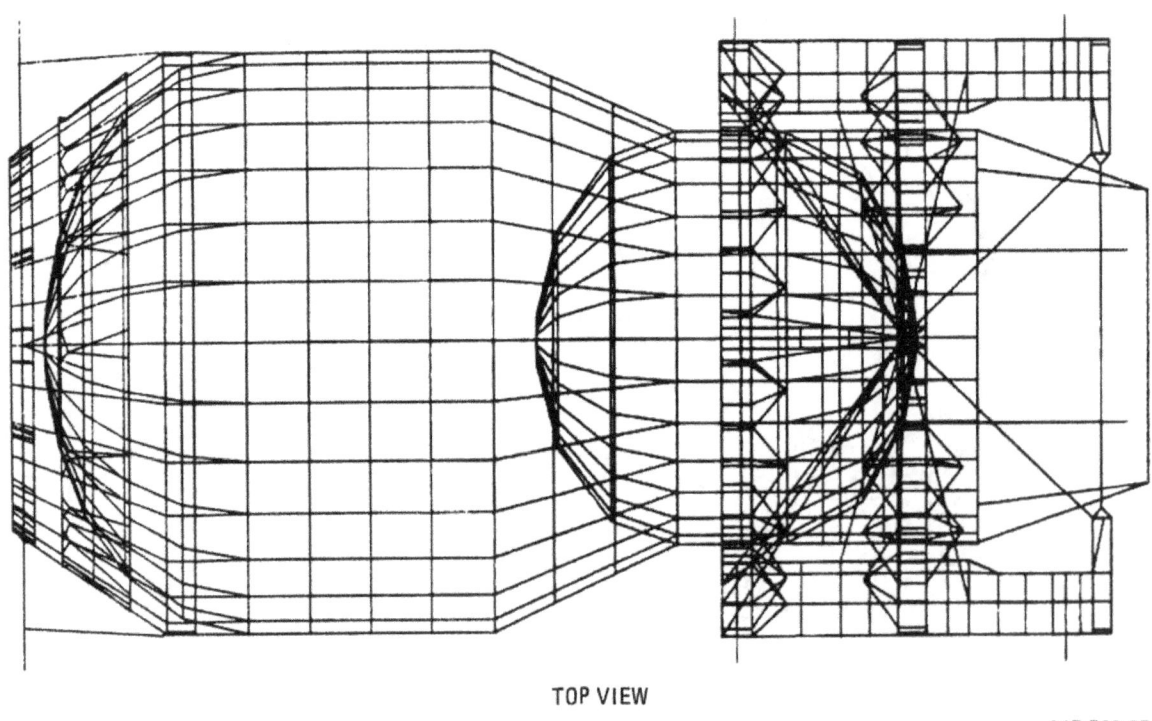

TOP VIEW

267.509-95

Figure 8-2.   Centaur G-Prime and CISS Finite Element Model

Structural flexibilities were then obtained from these detailed models and incorporated into a dynamics model of the entire Centaur G-prime. This dynamics model was then coupled, using modal synthesis, with STS dynamics models supplied by Rockwell International, and subjected to STS lift-off and landing forcing functions in dynamic transient response analyses. Cargo internal loads, accelerations, deflections, net load factors, and STS interface loads were calculated, as were spacecraft responses for the spacecraft contractors.

Quasi-static analyses were also performed in which the Centaur vehicle was coupled into the Orbiter flexibility matrix and subjected to quasistatic loads and Orbiter thermal deflections for ascent, on-orbit, and descent conditions. On-orbit thermal deflections are sufficiently small to avoid problems during a restow and relatch abort from the erected attitude.

The current Centaur G-prime is being analyzed with the 5302-pound Galileo and 809-pound International Solar Polar Mission (ISPM) ESA spacecraft.

All dynamic analyses to date have shown vehicle and spacecraft response accelerations and Orbiter interface loads to be less than those obtained using the preliminary design accelerations in ICD-2-1F001. The interface loads, in particular, show very significant reductions, and result in positive margins of safety for the Orbiter structure in all cases.

All structural requirements, such as factors of safety, fracture control, materials selection, etc., of JSC-07700, Vol. X Appendix 10.16 and ICD-2-1F001, will be adhered to in the Centaur G-prime design.

8.1.2 THERMODYNAMICS. The same heat transfer and thermodynamic design analysis methods developed for Centaur and refined by 20 years of application to Centaur missions are used. Centaur cargo element thermal analyses are being conducted to support design of Centaur G-prime vehicle and CISS structures, avionics, and fluid systems. Prelaunch and post-abort gas conditioning system analyses defines payload bay thermal environments. Thermal analyses of structures and structural assemblies provides thermal response data to support structural analysis and design, and yields main propellant tank heating rate partials to support fluid system analysis and design. Thermal analysis of components will support definition of component-level thermal control design requirements and qualification levels. A vehicle-level thermal math model has been formulated as input to a thermal integration analysis of the entire Centaur G-prime cargo element and will be used to provide refined definition of STS payload bay environment temperature levels.

Thermal design analyses of Centaur G-prime and CISS structures includes analysis of the entire forward structural assembly, including the forward adapter. An analysis of the aft structural assembly addresses the aft skirt, separation ring, deployment adapter, disconnect panels, and engine support structures. Detailed analysis of the CISS includes latching and support assemblies. Insulation blanket design analyses supports structural design goals, and also yields tank heating histories and purge requirements.

Fluids thermodynamic analyses are being conducted to support design of the
Centaur and CISS structures along with the fluid systems. Thermal analysis
support for fluid systems design includes a continuing update of main
propellant tank heating rates for all operational conditions, including
chilldown, tanking, in flight, and postabort. Additional thermal analysis in
support of fluids system designs includes tank pressurization, fill and drain,
reaction control, abort/dump, and main propulsion subsystems and components.
Thermodynamic analyses also support selection of pressurization and vent
systems, abort helium requirements, CISS helium storage bottle selection, and
mission propellant tank pressure profiles required to satisfy structural and
main engine NPSH requirements. Two areas of major emphasis for Centaur
G-prime development is the zero-g vent system and the abort dump system.

Zero-g vent capability is required to maintain Centaur G-prime hydrogen tank
pressure control while in the payload bay in low earth orbit and following
deployment from the Orbiter. GDC designed and tested prototype thermodynamic
vent systems in the 1960s and 1970s for liquid hydrogen and liquid oxygen
application. These systems demonstrated the feasibility of zero-g propellant
tank pressure control, which is achieved by throttling tank fluid through a
heat exchanger where it extracts energy from the bulk propellants. A
propellant mixing device, operating in the LO_2 tank, ensures thermal
equilibrum low-pressure rise rates to eliminate the need for on-orbit venting
the Centaur G-prime baseline one-burn missions.

The Centaur G-prime abort dump system is designed for compatibility with all
Shuttle abort modes that can occur before vehicle deployment. For these
aborts, methods are being devised for operating Centaur safely and for
disposing of propellants before landing. JSC has indicated dumping durations
as short as 250 seconds could be imposed during a return-to-launch site
abort. Experimentation described in Section 8.2 will provide empirical data
on the extent to which frozen equilibrium conditions exist during propellant
dump. Analysis of the abort dump procedure allows establishment of design
requirements for the propellant dump lines, pressurization technique, and tank
inerting. The abort dump flow-rate histories serves as an input to the
Rockwell International study for abort safety considerations.

8.1.3 PERFORMANCE. Trajectory design analyses for the reference missions are
being conducted to determine and define trajectory characteristics and
sequencing. Variations of performance and trajectory geometry are developed
for the missions, and nominal and dispersed closed-loop trajectories are
simulated and provided to the various design and analysis disciplines.
Centaur flight performance reserve requirements are determined using the Monte
Carlo technique. Trade studies are performed to maximize the performance
margin for nominal and backup deployment times. Factors affecting selection
of the nominal and contingency windows include performance capabilities of the
Orbiter and Centaur, Orbiter crew schedule constraints, tracking and telemetry
requirements of the Orbiter, Centaur, and spacecraft, and possible launch
delays.

8.1.4 GUIDANCE. A preliminary guidance system accuracy analysis for two Centaur G-prime planetary missions has been performed. The results show the expected requirements can be satisfied by the Centaur inertial guidance system and baseline avionics. GDC's analysis considered a navigation update using the Tracking and Data Relay Satellite System, rotation of the inertial platform to cancel the effects of gyro drifts, or an attitude update using a star scanner. The navigation and attitude updates were assumed 90 minutes before main engine start. Table 8-1 summarizes typical FOMs computed at injection plus ten days for the Galileo and ISPM missions.

Table 8-1. Typical FOMs (msec) for Galileo and ISPM Missions

Injection Rev.	Analysis Mode	FOM (msec)	
		Galileo 1986	IPSM 1986
5	No updates	21.3	29.1
	Platform rotation	10.9	15.0
	Attitude update	10.2	16.3
6	No update	25.8	36.2
	Platform rotation	14.8	18.3
	Attitude update	13.9	—
7	No update	31.5	41.7
	Platform rotation	22.9	23.6
	Attitude update	18.6	12.5
	Mission Requirements	24.0	35.0

8.1.5 STABILITY AND CONTROL. A dynamic model of Centaur, including elastic and slosh properties, has been furnished to evaluate the STS control capabilities and stability by coupling the Centaur and STS models.

Centaur erection, separation, and coast has been evaluated to verify full compliance with STS requirements, including component failure effects on stability and control. This analysis includes six-degree-of-freedom (6 DOF) simulations of both Centaur and STS to define the separation clearance envelope.

Centaur powered and coast phase control systems are being established and analyzed to assure vehicle stability and control. This will include the effects of the modified propellant tanks and revised vehicle stiffness. These analyses will include a time-domain 6 DOF simulation with all nonlinearities and time-varying properties modeled.

Modern control techniques will be employed to evaluate the noise and multiloop/multidisturbance effects. Frequency-domain analyses, such as root locus for basic stability and modified Z transform to evaluate the effect of sampling and digital computations, are performed. The effects of engine startup and shutdown on control will be delineated. A coast phase sequence is being generated and hydrazine usage determined. This sequence will be developed to satisfy the requirements for payload separation (e.g., spin-up, spin-down) and Centaur tank blowdown.

8.1.6 CISS CONTROL SYSTEM SAFETY. Verification of the integrity of the failure-tolerant CISS control system design will be accomplished by three analytical approaches.

1. An Interdata 8/32 realtime simulation is performed on the control system and the vehicle system being controlled (e.g., vehicle tank pressure). All control commands and responses are simulated to demonstrate the effects of multiple failures (e.g., open relays, failed valves) under worst-case variations of time in the five asynchronously operating control units.

2. Sneak circuit analysis is being conducted using NASA-developed pathfinding programs for analog circuitry and computer-aided digital analysis techniques for complex digital circuitry. This analysis identifies latent electrical circuit paths and conditions that can result in an undesirable event occurring without component failure.

3. A qualitative fault-tree analysis is being performed to examine the interactions between CISS and Centaur subsystems and to verify that overall system fault-tolerance levels have been achieved. The fault trees aid in identifying common cause and common mode failures that could defeat the intended level of failure tolerance.

8.2 TEST PROGRAM

Ongoing development, qualification, and validation tests being performed for Centaur G-prime and previous experience with cryogenic Atlas and Centaur vehicles will result in a low-risk program. All testing requirements for the Centaur G-prime program are included in the Integrated Test Plan which is a NASA/LeRC approved document.

8.2.1 DEVELOPMENT TESTS. Development testing for Centaur G-prime is planned to provide early solutions to design problems and to identify key characteristics of hardware and software. Component and/or subsystems will be tested in progressive stages to ensure earliest recognition of possible problem areas.

Development tests include all tests that are not qualification, acceptance, or prelaunch validation tests. The test program summary (Table 8-2) depicts test items and associated conditions. The test program flow diagram (Figure 8-3) shows the integrated test program flow, which describes an orderly progression to meet the program test objectives and test requirements. The test program is built on developmental testing experience of Centaur D-1A and the cruise missile.

Table 8-2. Test Program Summary

Test Item	Planning Test	Test Location	Hardware Configuration	Support Requirements
STRUCTURES				
1. Insulation System Material	Determine Electrostatic, Solar Emittance, Mechanical, & Outgassing Properties	GDC Material Research	Test Hardware — Various Samples of Insulation Material	Existing Test Facility & Support Test Equipment
2. Structural Test-Forward Adapter Support Struts	Demonstrate the Structural Strength & Stability of the Forward Support Struts. Verify Calculated Strength	GDC Material Research	Test Hardware — Forward Support Struts with End Fittings	Existing Test Facility & Support Test Equipment
3. Structural Test-CSS Structural Members	Demonstrate the Structural Strength & Stability of the Panel, Aft Support Bar & Pin	GDC Material Research	Test Hardware — Machined Panel, Aft Support Bar Segment with End Fitting & Machined Interface Pin	Existing Test Facility & Support Test Equipment
4. Structural Test-Spotweld/Seamweld Tank Joints	Perform Structural Test on Various Type of Joint Coupons of Tank Skins & Rings	GDC Material Research	Test Hardware — Specimens are Coupons made of Build-up on Skin/Doubler/Ring Thicknesses joined by Weld Like Production Article	Existing Test Facility & Support Test Equipment
5. Structural Test-CSS Gas Duct Bellows	Perform Load Deflection Test & Permanent Deflection Test on Gas Conditioning Bellows	GDC Material Research	Test Hardware — Bellows with End Fittings for Holding During Test	Existing Test Facility & Support Test Equipment
6. Centaur Functional-Deployable Antenna	Functional Test to Demonstrate Deployment of Antenna Support Structure	GDC Environmental Lab	Test Hardware — Spring Loaded Support Structure Released by Redundant Pyrotechnic Pin Pullers	Existing Test Facility & Support Test Equipment
7. CISS Functional-Centaur Severance System	Demonstrate the SUPER* ZIP Functions in Flight Environment (Primarily Low Temperature)	Vender LMSC Environmental Lab	Test Samples — SUPER* ZIP	Existing Test Facility & Support Test Equipment
8. ISPM Mission Peculiar Adapter Static Load Test	Demonstrate the Structural Integrity of the ISPM MPA	GDC Structural Lab	Test Hardware — Flight ISPM Mission Peculiar Adapter & Interfacing NASA Adapter	Existing Test Facility & Support Test Equipment
9. Centaur Cargo Element Dynamic Response Test	Determine Principal Mode Shapes and/or Responses for the CCE in the Vertical Position	GDC Sycamore Canyon Test Site	Test Hardware — Test CISS, Test Deployment Adapter, Test Centaur G', Test Forward Adapter, Aft P/L Adapter, Simulated S/C, STS Simulated Latches, Test Aft Adapter	Existing Test Facility & Support Test Equipment
10. Composite Panel Compression/Shear Load Test	Demonstrate Load Carrying Capability of the Composite Aft Adapter Panel. Loading of Panel Under Combined Compression and Shear Loading to Ultimate Load Then to Failure	Northrop, Hawthorne	Test Hardware — Aft Adapter Composite Cut Out Panel	Existing Test Facility & Support Test Equipment

267.509-99-1

Table 8-2. Test Program Summary, Contd

Test Item	Planning Test	Test Location	Hardware Configuration	Support Requirements
11. Composite Adapter Panel Thermal Gradient Compression Test	Obtain Load Deflection Data on Composite Adapter Panels While Maintaining a Thermal Differential Along the Length of the Panel	GDC M&P Lab	Test Hardware — Forward Adapter Panel, Aft Adapter Panel	Existing Test Facility & Support Test Equipment
12. CISS Functional Spring Thrust System	Demonstrate that Spring Thrust System will Provide Energy to Thrust Centaur Vehicle from Deployment Adapter with Required Acceleration/Velocity	GDC KM Test Tower	Test Hardware — Spring Thrust System, Engine Support Cone, Electrical Flyaway Umbilicals, Fluid System, Disconnects, Deployment Adapter & SUPER* ZIP Ring	Existing Test Facility & Support Test Equipment
13. CSS Stiffness Test	Demonstrate that CSS Exhibits Satisfactory Structural Properties when Subjected to Ultimate Design Loads	GDC Structural Lab	Test Hardware — Centaur Support Structural	Existing Test Facility & Support Test Equipment
14. Centaur Structure & CSS Static Load & Stiffness Test	Demonstrate the Structural Integrity & Stiffness of the Centaur Tank, Forward & Aft Adapter Components & CISS	GDC Sycamore Canyon Test Site	Test Hardware — Aft Payload Adapter, Forward Adapter, Centaur Test Vehicle, Aft Adapter, Deployment Adapter Including Engine Support Structure and CSS. Tank must be Complete Enough for Cryo Tanking	Existing Test Facility — New Support Structure
15. Centaur LO_2 Tank Aft Bulkhead Hydrodynamics Loads & Deflection	Validate LO_2 Aft Bulkhead for Application of Longitudinal & Lateral Dynamic. Hydrodynamics Pressure Loads at STS Liftoff	GDC CEVAT Centrifuge	Test Hardware — Existing Stub Tank	Existing Test Facility & Support Test Equipment
16. LH_2 Tank Sidewall Blanket	Evaluate Tank Thermal Properties During Pretanking, Chilldown, & Tanking	GDC Sycamore Canyon Test Site	Test Hardware — Test Centaur Tank with PLIS Probes, All Forward Adapters, Blanket System, Purge System, Vent System, Aft Adapter, & CSS	Existing Test Facility — New Support Structure Required
17. Centaur Tank, Adapters & CISS Dynamic Modal Survey	Determine Principal Mode Shapes & Natural Frequency of the Centaur/CISS	GDC KM Vibration Tower	Test Hardware — Complete Test Centaur Tank, Adapters, Test CISS & Simulated Payload	Existing Test Facility & Support Test Equipment
18. Deployment Adapter Acoustic Test	Measure Environmental Vibration on Simulated Components Attached to Deployment Adapter with Shuttle Acoustic Levels	Goddard Space Flight Center	Test Hardware — Aft Deployment Adapter & Set of Simulated Components	Existing Test Facility & Support Test Equipment

267.509-99-2

Table 8-2. Test Program Summary, Contd

Test Item	Planning Test	Test Location	Hardware Configuration	Support Requirements
19. Forward Adapter Acoustic Test	Measure Environmental Vibration on Simulated Components Attached to Forward Adapter with Shuttle Acoustic Levels	Goddard Space Flight Center	Test Hardware — Production Forward Adapter, Stub Adapter & Set of Simulated Avionics	Existing Test Facility & Support Test Equipment
FLUIDS				
20. Cryo Flange	GHe & Cryogenic Leakage of Various Flange/Joint Sealing Methods Used on Centaur — CISS	GDC Sycamore Canyon Test Site	Test Hardware — Flanges/Joints Configured per Centaur — CISS Application	Existing Test Facility & Support Test Equipment
21. Elemental Mechanisms	Early Demonstration of Critical Elements of the Mechanical Systems	GDC KM Test Tower	Test Hardware — Disconnects, Seals & Materials	Existing Test Facility & Support Test Equipment
22. Ascent Vent Valve-LH$_2$	Determine Pressure Drop Data at Low Vent Rates	GDC KM Fluids Lab	Test Hardware — Flight Configuration Valve	Existing Test Facility & Support Test Equipment
23. Insulation Blanket Purge System	Demonstrate Purge System Capability to Maintain Insulation Blanket Delta-P Under Various Conditions	GDC Sycamore Canyon Test Site	Test Hardware — Flight-type Solenoid Valves, Delta-P Transducers, Avionics, Test Box, Orifices, Insulation Relief Valves & Regulator	Existing Test Facility — New Support Structure Required
24. LO$_2$ Jet Mixer	Verify Jet Mixer Provides Proper Propellant Mixing in the LO$_2$ Tank & Components Perform Properly as a System	GDC KM Fluids Lab	Test Hardware — 0.2 Scale System Hardware	Existing Test Facility — Support Test Equipment
25. Tank Two-Phase Outflow Test	Perform Outflow Test with Scaled Down LH$_2$ & LO$_2$ Tanks to Determine Quantity of Liquid Being Expelled as a Function of Tank Residuals	GDC Sycamore Canyon Test Site	Test Hardware — 1/5 Scale & 1/20 Scale Tanks with Simulated LO$_2$ & LH$_2$ Tank Outlets	Existing Test Facility — New Support Structure Required
26. Dump System Flow Test	Conduct Individual LN$_2$ & LH$_2$ Flow Tests Through a Representative Flight Propellant Dump System to Verify Computer Model	GDC Sycamore Canyon Test Site	Test Hardware — 0.4 Scale Model of Propellant Dump System	Existing Test Facility — New Support Structure Required
27. Centaur Rotation System Operation	Functional Systems Tests Include Rotation Followed by Deployment Adapter Retraction, Also Demonstrate the Backup Rotation System	GDC KM Test Tower	Test Hardware — Oxidizer Gimbal Ducts, Flex Hoses, D&A Rotation Systems, (CISS & D/A)	Existing Test Facility & Support Test Equipment

267.509-99-3

Table 8-2. Test Program Summary, Contd

Test Item	Planning Test	Test Location	Hardware/Configuration	Support Requirements
28. Fill/Drain Residual Purge	Verify the Feasibility of Reducing Pre-liftoff Residuals to an Acceptable Level by Purging the LO$_2$ & LH$_2$ Fill/Drain Lines	GDC KM Fluids Lab	Test Hardware — Tygon Tubing Simulation of the Fill/Drain Lines	Existing Test Facility & Support Test Equipment
29. LH$_2$ Propellant Mixing	Verify Zero-g Vent Mixer Will Provide Proper Propellant Mixing	GDC KM Fluids Lab	Test Hardware — 0.2 and 0.05 Scale System Hardware	Existing Test Facility & Support Test Equipment
30. Engine Actuator Stiffness	Determine Stiffness & Damping Characteristics of Actuator in 5 to 50 Hz Range	GDC KM Vibration Lab	Test Hardware — Engine Actuator & Servo Valve	Existing Test Facility & Support Test Equipment
31. LH$_2$ Sump Design	Perform LH$_2$ Outflow Test With Simulated Scale Model Tank to Determine Sump Design	GDC Sycamore Test Site	Test Hardware — Simplified Scale Model of LH$_2$ Tank With LH$_2$ Tank Outlets	Existing Test Facility & Support Test Equipment
32. Zero-G Heat Exchanger	Evaluate Zero-g Heat Exchanger	Vendor Geoscience	Test Hardware — Prototype Heat Exchange	Existing Test Facility & Support Test Equipment
33. High-Angulation Gimballed Joint	Verify Feasibility of Three-Gimbal Joint Ducting System	Vendor Ametek, Straza Div	Test Hardware — Vacuum Jacketed Gimballed Joint	Existing Test Facility & Support Test Equipment
34. LH$_2$ TVS Components	Determine Heat Exchanger Performance During Ground Test in LH$_2$	GDC Sycamore Canyon Test Site	Heat Exchanger With Lines & Fittings	Existing Test Facility & Support Test Equipment
35. Engine Control System	Define Helium Control Pressure Available to Main Engine Prestart Solenoid Valves During Engine Start	GDC KM Fluids Lab	Test Hardware — Two Helium Control Systems	Existing Test Facility & Support Test Equipment
36. Hydrazine System Ground Energized Heaters	Evaluate Ground Powered Heaters to Maintain Hydrazine Fluid	GDC KM Fluid Lab	Test Hardware — Hydrazine Tank and Heater	Existing Test Facility & Support Test Equipment
37. Lipseal/Rosan Adapter Connectors	Evaluate Connectors for Pneumatic System Operation	GDC Sycamore Canyon Test Site	Test Hardware — Tubes & Lipseal Connectors	Existing Test Facility & Support Test Equipment
38. Relief Valve	Perform a Burst Test on Relief Valve	GDC Sycamore Canyon Test Site	Test Hardware — Relief Valve, Pressure Gage	Existing Test Facility & Support Test Equipment

267.509-99-4

Table 8-2. Test Program Summary, Contd

Test Item	Planning Test	Test Location	Hardware Configuration	Support Requirements
39. LH$_2$ On Orbit Abort Dump Pullthrough	Determine Vapor Pull-through Height, Rate	GDC Sycamore Canyon Test Site	Plexiglass Model Tank Flowmeter, Camera	Existing Test Facility & Support Test Equipment
40. LO$_2$ On Orbit Abort Dump Pullthrough	Determine Vapor Pull-through Height Rate	GDC Sycamore Canyon Test Site	Test Hardware – Plexiglass Model Tank Flowmeter, Camera	Existing Test Facility & Support Test Equipment
41. CISS LO$_2$ & GO$_2$ Line Assembly Vibration	Verify Ducting and Support Structure Design	GDC KM Vibration Lab	Test Hardware – One Line Assembly	Existing Test Facility & Support Test Equipment
42. CISS LH$_2$ & GH$_2$ Line Assembly Vibration	Verify Ducting and Support Structure Design	GDC KM Vibration Lab	Test Hardware – One Line Assembly	Existing Test Facility & Support Test Equipment
43. RCS Line & REM Heaters	Verify Heaters Performance in Deep Space Environment	GDC KM Fluid Lab	Test Hardware – One set of System Heaters	Existing Test Facility & Support Test Equipment
44. High Pressure Tubing Buttwelded Burst	Determine Burst Pressure of Each Weld Specimen	GDC KM Fluids Lab	Test Hardware – Three of each Size (1/4, 3/8, 1/2 and 1)	Existing Test Facility & Support Test Equipment
AVIONICS				
45. LH$_2$ PLIS Probe	Evaluate LH$_2$ PLIS Probe	GDC KM Vibration Lab	Test Hardware – LH$_2$ PLIS Probe	Existing Test Facility & Support Test Equipment
46. LO$_2$ PLIS Probe	Evaluate LO$_2$ PLIS Probe	GDC KM Vibration Lab	Test Hardware – LO$_2$ PLIS Probe	Existing Test Facility & Support Test Equipment
47. LH$_2$ PU Probe	Evaluate LH$_2$ PU Probe & Harness	GDC KM Vibration Lab	Test Hardware – LH$_2$ PU Probe	Existing Test Facility & Support Test Equipment
48. LO$_2$ PU Probe	Evaluate LO$_2$ PU Probe & Harness	GDC KM Vibration Lab	Test Hardware – LO$_2$ PU Probe	Existing Test Facility & Support Test Equipment
49. CISS Battery Performance	Demonstrate Current & Voltage Performance both Parallel & Single Battery. Demonstrate Single Battery Capacity Under Worst Case Loads	GDC KM Battery Lab	Test Hardware – Eagle Picher Battery	Existing Test Facility & Support Test Equipment
50. TLM Coax Cable Assemblies	Evaluate TLM Coax Cable Assemblies in Conjunction with Telemetry Subsystem Compatibility Test	GDC KM RF Lab	Test Hardware – Coax Cable Assemblies	Existing Test Facility & Support Test Equipment

267.509-99-5

Table 8-2. Test Program Summary, Contd

Test Item	Planning Test	Test Location	Hardware Configuration	Support Requirements
51. R.F. Antenna Patterns	Perform RF Radiation Pattern Testing with Antennas Mounted on 1/4 Scale Centaur Mockup	GDE KM Antenna Range	Test Hardware — Centaur Mockup (1/4 Scale), Two Antennas (1/4 Scale), Spacecraft Mockup (1/4 Scale)	Existing Test Facility & Support Test Equipment
52. Dual Failure Tolerant Arm/Safe Sequencer	Demonstrate Functionally Compatible During Thermal Vacuum, Shock & Vibration Testing	GDC Environmental Lab	Test Hardware — Prototype DUFTAS	Existing Test Facility & Support Test Equipment
53. Control Unit Performance	Demonstrate Functionally Compatible during Thermal Vacuum, Shock, Vibration Testing	GDC KM SIL & Environmental Lab	Test Hardware — Prototype CUs & Replica Unit	Existing Test Facility & Support Test Equipment
54. Control Distribution Unit	Demonstrate Functionally Compatible during Temperature, Vibration & Shock Testing	GDC KM SIL & Environmental Lab	Test Hardware — Replica Unit	Existing Test Facility & Support Test Equipment
55. PLIU Performance	Demonstrate Functionally Compatible During Temperature, Vibration & Shock Testing	GDC KM SIL & Environmental Lab	Test Hardware — Replica Unit	Existing Test Facility & Support Test Equipment
56. Pyro Initiator Controller	Demonstrate Pyro Initiator Controller Capability during. Thermal Vac, Shock, and Vibration Testing	GDC KM Environmental Lab	Test Hardware — Replica Pyro Initiator Controller	Existing Test Facility & Support Test Equipment
57. Electrical Distribution Unit	Demonstrate Functionally Compatible during Thermal Vacuum, Shock & Vibration Testing	GDC KM Mechanical Lab	Test Hardware — Prototype	Existing Test Facility & Support Test Equipment
58. Vehicle DCU Software-Pre-Flt Prog. Module Level & Integrated Level	Test All Preflight (Pre-Liftoff) Modules Separately & then together as an Integrated Program	GDC KM TICS & SIL Labs	Test Hardware — N/A	Existing Test Facility & Support Test Equipment
59. Vehicle DCU Software-Flt. Prog. Module Level & Integrated Level	Test All Flight Modules Separately & then Together as an Integrated Program	GDC KM "TICS/TRAJEX" on GPC & "DFAST" on Harris	Test Hardware — N/A	Existing Test Facility & Support Test Equipment
60. CISS DCU & CU Software-Prog. Module Level & Integrated Level	Test All Software Modules Separately & then Together as an Integrated Program	GDC KM SIL Lab "CFAST" on Harris	Test Hardware — N/A	Existing Test Facility & Support Test Equipment
61. CCLS Software	CCLS Software Development, Integration & Validation Testing	GDC KM SIL Lab	Test Hardware — N/A	Existing Test Facility & Support Test Equipment

267.509-99-6

Table 8-2. Test Program Summary, Contd

Test Item	Planning Test	Test Location	Hardware Configuration	Support Requirements
62. Guidance & Navigation System	Verify Guidance/Navigation, Gyro Compassing & Star Scanner Hardware/Software Interfaces for New Modes of Operation	GDC KM SIL Lab	Test Hardware — Prototype DCU, IMG, & Engineering Unit	Existing Test Facility & Support Test Equipment
63. Avionics System Integration	System Level Hardware & Software Interface Com-patibility, Software & System Level Safety Demonstration	GDC KM SIL Lab	Test Hardware — Prototype CISS & Centaur Packages Simulators	Existing Test Facility & Support Test Equipment
64. EMI Evaluation	EMI Evaluation Testing on Centaur & CISS Avionics Boxes & System	GDC KM SIL Lab	Test Hardware — Prototype Centaur & CISS Units	Existing Test Facility & Support Test Equipment
65. EMC System Level	System Level EMC Testing on Centaur Vehicle & CISS	GDC Vehicle Final Assembly	Test Hardware Production Centaur & CISS	Existing Test Facility & Support Test Equipment
66. TLM Subsystem Integration	Perform Evaluation Tests for TLM Subsystem	GDC KM RF Lab & SIL	Test Hardware Prototype TLM Subsystem	Existing Test Facility & Support Test Equipment
67. TDRSS Compatibility Test	Verify Measurement of System Parameters, Demonstrate Compatibility With TDRSS	GDC GSFC CTV	Test Hardware — CTV GDC SIL	
68. Telemetry System Electronic System Test Laboratory Test	Demonstrate Compatibility Between Centaur TLM and the Orbiter RF Systems	JSC	Test Hardware — S-Band Transmitter, RF Amplifier, Switch, Coaxial Cable Assemblies	Existing Test Facility & Support Test Equipment
69. CISS Elect. Power System Performance Verification	Demonstrate Power System Capabilities	GDC KM SIL Lab	Test Hardware — 2 CISS Main Batteries	Existing Test Facility & Support Test Equipment
70. Sequence Control Unit	Demonstrate performance of Modified SCU During Vibration and Temperature	GDC Environ-mental Lab	Test Hardware — Prototype SCU	Existing Test Facility & Support Test Equipment
71. Uplink Downlink Unit	Evaluate UDU to Withstand Vibration, Shock, Thermal Vacuum	GDC Environ-mental Lab	Test Hardware — Box with Components and Connec-tors Mounted	Existing Test Facility & Support Test Equipment
72. Servo Inverter Unit	Demonstrate Performance of Modified SIU During Temperature and Vibration Testing	GDC Environ-mental Lab	Test Hardware — Prototype SIU	Existing Test Facility & Support Test Equipment
73. Instrumentation Box Evaluation	Demonstrate Structural Design during Vibration	GDC Environ-mental Lab	Test Hardware — Prototype Forward Instrumentation Box	

267.509-99-7

8-13

Table 8-2. Test Program Summary, Contd

Test Item	Planning Test	Test Location	Hardware Configuration	Support Requirements
GRD AVIONICS				
74. Antenna Test Couplers	Evaluate Antenna Test Couplers	GDC KM RF Lab	Test Hardware — Prototype Antenna Test Coupler, Prototype Antenna	Existing Test Facility & Support Test Equipment
75. GSE	Development Tests for: CCLS, Hardware Extension Remote, Missile Checkout Computer System, Software Development Computer System & Mobile Support Equipment	GDC KM SIL Lab	Test Hardware — CCLS, HER, MCCS, SDCS, & MSE	Existing Test Facility & Support Test Equipment
INTEGRATION TESTS				
76. Centaur/CISS/ Orbiter Mate Test at OPF	Demonstrate Compatibility of Centaur CISS Attach Fittings & Hardware Interfaces	KSC OPF	Test Hardware — Orbiter, Centaur & CISS	Existing KSC Facility, Support Test Equipment & Strongback
77. Terminal Countdown Demonstration (TCD) — Tanking Test at 36A	Chilldown, Fill to Flight Level, Drain, Dump System Validation	GDC ELS CX-36	Test Hardware — Centaur, CISS, Simulated Orbiter, CCLS, MSE & Prop Skids	Existing Test Facility. New Support Test Equipment
78. Fluid System Dry Mode Integration Test at CX-36	Perform Fluid System Dry Mode Integration Test before Flowing Cyrogenics or Helium for Three Configurations	GDC ELS CX-36	Test Hardware — LO$_2$/ LH$_2$/GHe Ground Systems, MSE, CCLS, CISS, MSE, Centaur	Existing Test Facility. New Support Test Equipment
79. LO$_2$/LH$_2$ Cold Flow Validation Tests at CX36	Perform Cold Flow Validation Tests for LO$_2$ & LH$_2$ Systems both before & After CISS Erection	GDC ELS CX-36	Test Hardware — LO$_2$/ LH$_2$ Ground Systems, MSE, CCLS, CISS, & MSE	Existing Test Facility. New Support Test Equipment
80. Integrated LO$_2$/LH$_2$ Cold Flow Validation Tests at CX-39	Perform One Time LO$_2$ & LH$_2$ Cold Flow Validation Tests prior to Launch Countdown Demonstration	KSC CX-39 A/B	Test Hardware — LO$_2$/LH$_2$ Control Skids & MSE Simulator Installed at CX-39 A&B	Existing KSC Facility. Modified Support Test Equipment
81. Integrated Launch Countdown Demonstration at CX-39	Prior to Launch Tanking Perform One-Time Tanking Test of Shuttle/Centaur	KSC CX-39 A/B	Test Hardware — Centaur & CISS, LH$_2$, LO$_2$ & GHe Control Skids Supporting Avionics & Orbiter	Existing KSC Facility, Modified Support Test Equipment
82. Centaur/Spacecraft Interface Test at VPF	Demonstrate Interface Compatibility of Centaur to Spacecraft	KSC — Vertical Processing Facility (VPF)	Test Hardware — Flight Hardware or Simulator	Existing KSC Facility & Support Test Equipment
83. Centaur/Spacecraft/TDRSS/MCC End to End Test at VFP	Verify Uplink/Downlink Data Thru Spacecraft, Centaur, Orbiter & JPL-MCC	KSC-VPF	Test Hardware — Flight Hardware or Simulator	Existing KSC Facility & Support Test Equipment

267.509-99-8

Table 8-2. Test Program Summary, Contd

Test Item	Planning Test	Test Location	Hardware Configuration	Support Requirements
84. Spacecraft/Centaur/Orbiter Functional Interface Test at VPF	Demonstrate Interface Compatibility Between Spacecraft/Centaur to Orbiter Avionics	KSC-VPF	Test Hardware — Flight Hardware or Simulator	Existing KSC Facility & Support Test Equipment
85. Mission Simulator Test at VPF	Perform Mission Simulation Sequence from Pre-Liftoff Thru Spacecraft Separation	KSC — VPF (with CITE)	Test Hardware — Flight Hardware & Software	Existing KSC Facility & Support Test Equipment
86. ECS Validation Test at CX-36	Perform Validation Tests on CX-36 Environmental Control System both Before & After Centaur Installation	GDC ELS CX-36	Test Hardware — CX-36 Air Conditioning Units, Simulated Orbiter Bay Enclosure & Interconnecting Ducting Plus Centaur	Existing Test Facility. Modified Support Test Equipment
87. Helium System Validation at CX-36	Perform Validation Tests on Helium Skid & Piping Both Before & After CISS Erection	GDC ELS CX-36	Test Hardware — GHe Ground System, MSE, CISS & CCLS	Existing Test Facility. Modified Support Test Equipment
88. Avionics Validation at CX-36	Demonstrate Centaur/CISS Avionics Compatibility with CX-36 GSE Prior to Flight Hardware	GDC ELS CX-36	Test Hardware — Flight Hardware Similator, CCLS & MSE	Existing Test Facility. Modified Support Test Equipment
89. Electrical/ Electronic Simulator Operations at VPF	Demonstrate Centaur/S/C/ Orbiter Avionics Interface Compatibility with VPF Modifications	KSC-VPF	Test Hardware — Flight Hardware Simulator, CCLS & MSE	Existing KSC Facility, Modified Support Test Equipment
90. Electrical/ Electronic Simulator Operations at CX-39	Demonstrate Compatibility with CX-39 Data and Control Circuits	KSC CX-39 A/B	Test Hardware — Flight Hardware Simulator, CCLS and MSE	Existing KSC Facility, Modified Support Test Equipment
91. Telemetry Parameter Measurements	Demonstrate Centaur TLM System Compatibility, with CX-36 & VPF GSE prior to Flight Hardware	GDC ELS CX-36 & KSC-VPF	Test Hardware — Flight Hardware Simulator	Existing Test Facility, Modified Support Test Equipment
92. Centaur/Tracking & Data Relay Satellite System (TDRSS) Compatibility Facility	Demonstrate Compatibility with the TDRSS Using the Direct Relay Link Through Merritt Island Launch Area (MILA) Prior to Flight Hardware	VPF & MILA & GDC ELS at CX-36	Test Hardware — Flight Hardware Simulator	Existing Test Facility, Modified Support Test Equipment
93. Centaur Tracking & Data Relay Satellite System (TDRSS) Compatibility Test	Demonstrate Compatibility with the TDRSS Using the Direct Relay Link Through Merritt Island Launch Area (MILA) Prior to Flight Hardware	VPF & MILA & GDC ELS at VPF	Test Hardware — Flight Hardware Simulator	Existing Test Facility, Modified Support Test Equipment

267.509-99-9

Table 8-2. Test Program Summary, Contd

Test Item	Planning Test	Test Location	Hardware Configuration	Support Requirements
94. Centaur/Orbiter RF Link Demonstration	Perform Closed Loop RF Xmission from Centaur to Orbiter Payload Interrogator in VPF Prior to Flight Hardware. Process Data Through Orbiter & Record Data at Site	KSC-VPF	Test Hardware Flight Hardware Simulator	Existing Test Facility, Modified Support Test Equipment

267.509-99-10

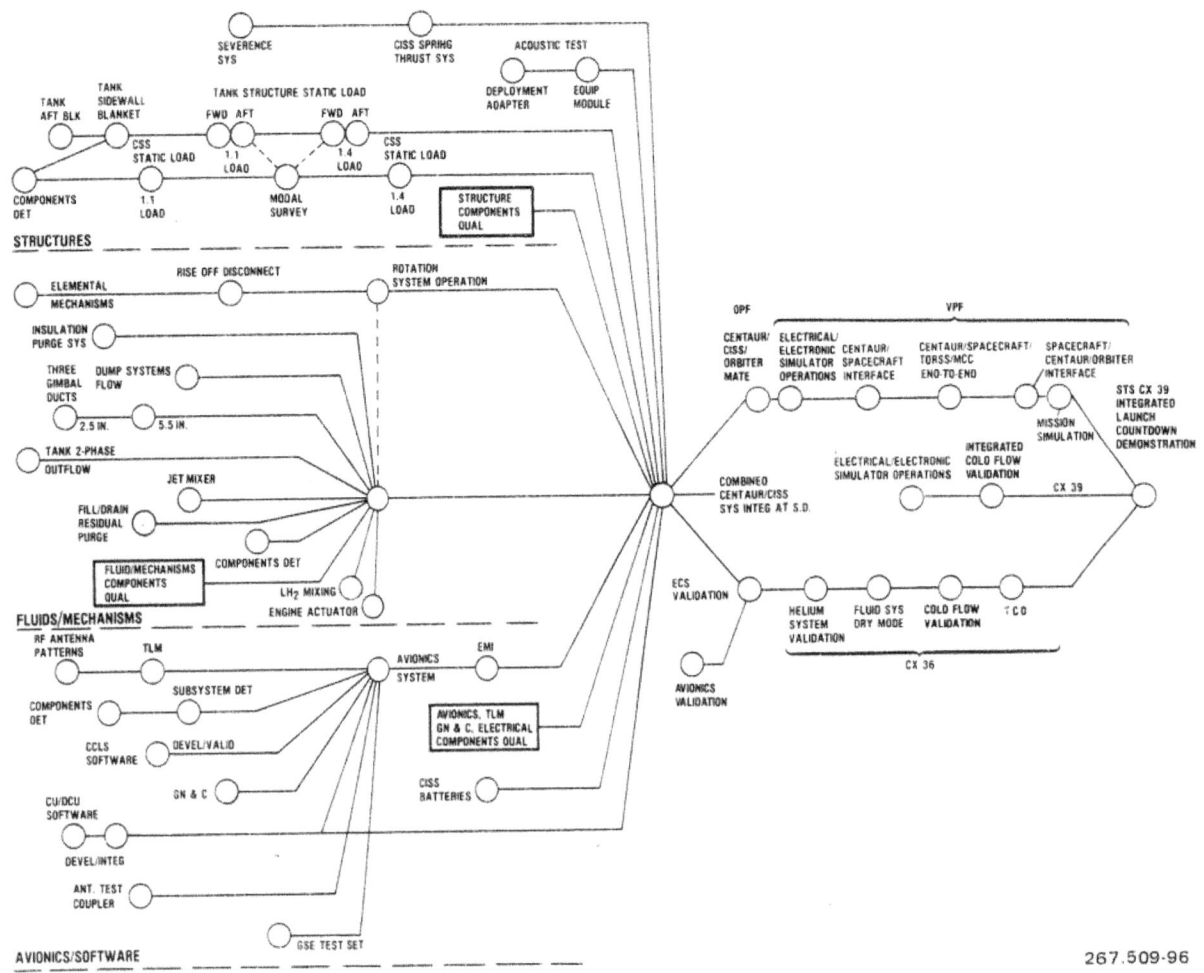

267.509-96

8.2.1.1 <u>Structures</u>. A limited amount of structural development testing will be necessary for Centaur G-prime. These development tests will consist initially of material elements and welding tests, then separation tests, tank insulation tests, Centaur support structure load tests, and tests due to the new launch and landing environment. These new environment(s) tests include acoustics, vibration, Centaur loads, and modal survey.

8.2.1.2 <u>Fluids and Mechanisms</u>. Fluids and mechanisms systems for Centaur G-prime are based on Centaur D-1A technology except where Shuttle imposes additional requirements. The development test will cover mechanisms to deploy the Centaur from the Orbiter, propellant dump capabilities, and fluid interfaces with CISS.

8.2.1.3 <u>Avionics and Software</u>. Avionics and software on the Centaur G-prime evolved from Centaur D-1A systems with all changes required because of the additional Shuttle requirements such as safety, operations sequence, different-shaped vehicle, and platform torquing functions.

8.2.1.4 <u>Ground Avionics</u>. Although most ground avionics will be the same as used on D-1A Centaur, a few additional items of test equipment or test sets will be required to test the Centaur G-prime avionics and software.

8.2.1.5 <u>CISS/Centaur Integrated Tests</u>. The functional integration tests verify the Centaur vehicle with CISS will meet vehicle-level performance requirements. The initial phase of functional integration consists of test on each functional subsystem using the other systems as necessary to support the tests. These individual subsystem tests will concentrate on detailed performance and margins of each individual subsystem. The test flow is shown in Figure 8-4.

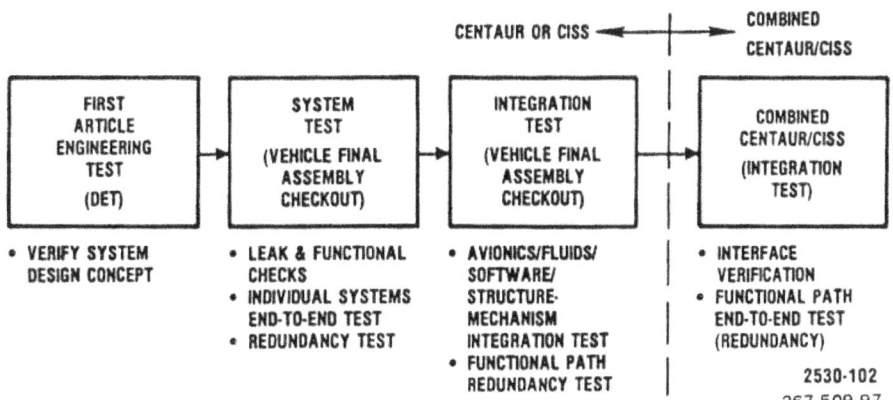

Figure 8-4. CISS/Centaur Test Flow

Factory testing of the Centaur first article will be similar to current practice. New tasks include checkout of redundant systems, Centaur-to-CISS interfaces, and combined Centaur/CISS system tests. Since CISS is a new item, the test philosophy will be similar to Centaur testing using the CISS simulator.

8.2.2 COMPONENT QUALIFICATION TESTS. Early qualification testing will reduce cost and increase reliability. Component qualification testing for Centaur G-prime will discover any problems before system and subsystem-level testing.

All components will have successfully completed functional checkout and acceptance testing including burn-in (if required) before qualification testing. Environmental qualification test requirements will comply with ICD-2-1F001 as interpreted by 65-00206B. All newly designed components will be qualified to ensure full compliance with Shuttle requirements. New components that require qualification testing include six structures, 49 fluids and mechanisms, and 24 avionics.

8.2.3 VALIDATION TESTS. Validation tests on Centaur G-prime will increase reliability. These tests will be performed at the launch site and are depicted in Table 8-2.

SECTION 9

NEW CONTENT

9.1.1   INTRODUCTION

This chapter was not included in the original report; it is Bonus
Material (New Content). Section 9 comprises:

*   Figures:

     Many of the figures present in the original document had very
     small text and/or low contrast. This chapter includes enlarged
     and enhanced versions of these figures.

*   Photographs:

     Several photographs, the original color versions of which are all
     courtesy NASA, from the history of Shuttle-Centaur are included.
     These include images of Shuttle-Centaur scale models, and the
     rollout ceremony for the first Shuttle-Centaur stage.

*   Other Shuttle-Centaur Reports:

     Two contemporary NASA reports are also reproduced in Section 9:
     "Centaur for the 1980s" (1981) and "A High Energy Stage for the
     National Space Transportation System" (1984).

9.1.2   COPYRIGHT NOTICE

9.1.2.1   Public Domain

The original versions of the figures, photographs, images and text in
Section 9 are believed to be in the public domain.

9.1.2.2   Derivative Works

All of the figures, photographs, images and text in Section 9 have been
enlarged and enhanced using image editing software (Photoshop®, Nik
SilverFx). Each page from the original pdf downloads was saved as a
Photoshop® file, wherein adjustments were made to (1) Enhance their
clarity and legibility (e.g., Contrast, Levels and/or Exposure layers
were made and adjusted); (2) Correct their orientation (e.g., tilted
pages and mirror images were corrected; (3) Remove minor defects (e.g.,
images of staples or binding rings are removed); (4) Adjust their size
and border (i.e., they were cropped).

Each individual page from the original report was used to create a
unique image file that was individually enhanced in this manner. The
enhanced versions of these images are derivative works that are ©2014
Mooncat Publications. Section 9 is ©2014 Mooncat Publications.

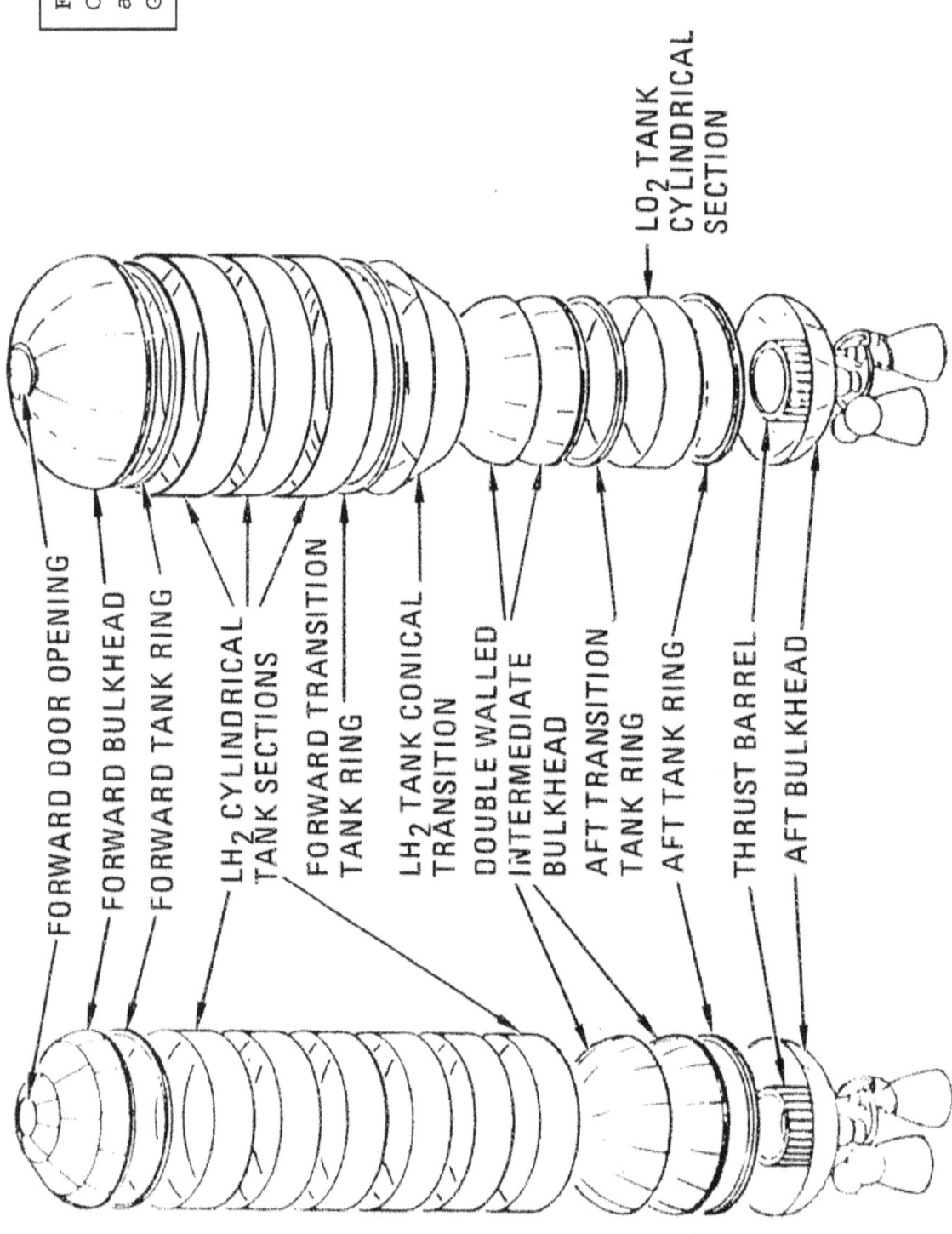

Fig. 3-5
Centaur D-1A
and Centaur
G-Prime Tanks

267.509-19

FORWARD DOOR OPENING

FORWARD BULKHEAD

FORWARD TANK RING

LH₂ CYLINDRICAL TANK SECTIONS

FORWARD TRANSITION TANK RING

LH₂ TANK CONICAL TRANSITION

DOUBLE WALLED INTERMEDIATE BULKHEAD

AFT TRANSITION TANK RING

AFT TANK RING

THRUST BARREL

AFT BULKHEAD

LO₂ TANK CYLINDRICAL SECTION

CENTAUR G-PRIME

CENTAUR D-1A

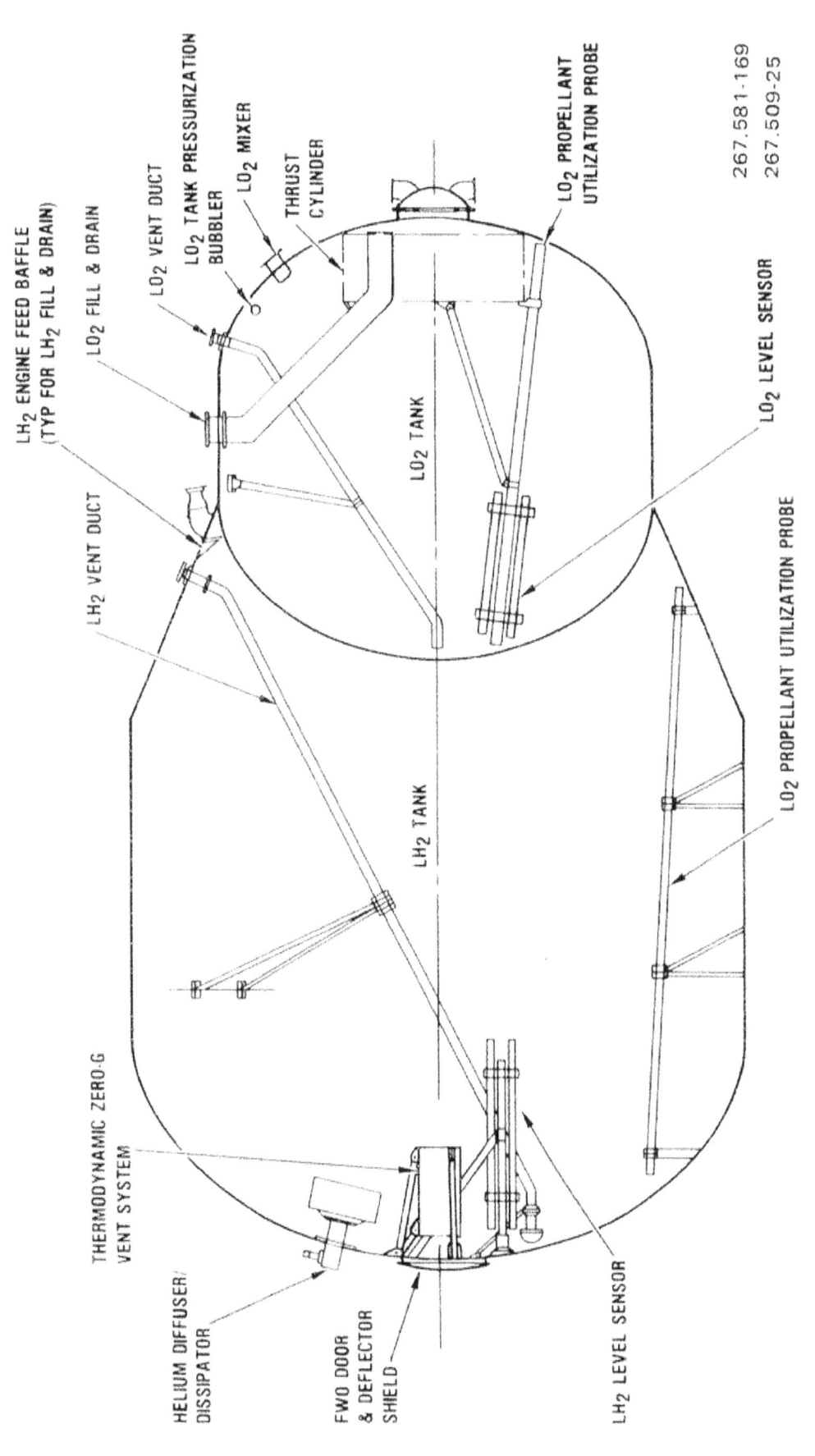

LH₂ ENGINE FEED BAFFLE
(TYP FOR LH₂ FILL & DRAIN)

LO₂ FILL & DRAIN

LO₂ VENT DUCT

LO₂ TANK PRESSURIZATION BUBBLER

LO₂ MIXER

THRUST CYLINDER

LO₂ PROPELLANT UTILIZATION PROBE

267.581-169
267.509-25

LO₂ TANK

LO₂ LEVEL SENSOR

LH₂ VENT DUCT

LO₂ PROPELLANT UTILIZATION PROBE

LH₂ TANK

THERMODYNAMIC ZERO-G VENT SYSTEM

HELIUM DIFFUSER/DISSIPATOR

FWD DOOR & DEFLECTOR SHIELD

LH₂ LEVEL SENSOR

Figure 3-11. Centaur G-Prime Tank Internal Installations

9-3

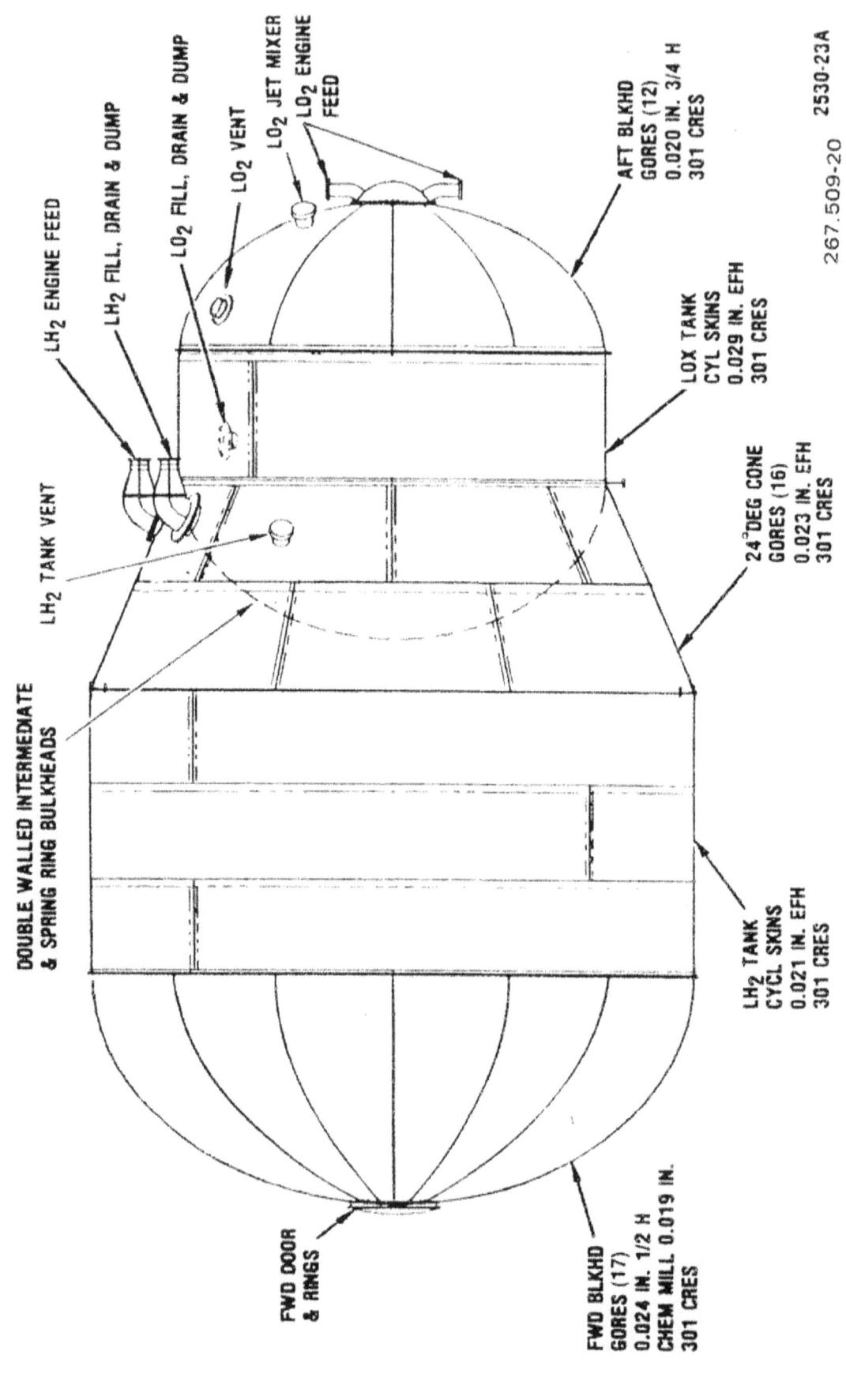

DOUBLE WALLED INTERMEDIATE & SPRING RING BULKHEADS

LH₂ TANK VENT

LH₂ ENGINE FEED

LH₂ FILL, DRAIN & DUMP

LO₂ FILL, DRAIN & DUMP

LO₂ VENT

LO₂ JET MIXER

LO₂ ENGINE FEED

AFT BLKHD GORES (12) 0.020 IN. 3/4 H 301 CRES

267.509-20   2530-23A

LOX TANK CYL SKINS 0.029 IN. EFH 301 CRES

24°DEG CONE GORES (16) 0.023 IN. EFH 301 CRES

LH₂ TANK CYCL SKINS 0.021 IN. EFH 301 CRES

FWD DOOR & RINGS

FWD BLKHD GORES (17) 0.024 IN. 1/2 H CHEM MILL 0.019 IN. 301 CRES

Figure 3-6. Centaur G-Prime Propellant Tank Configuration

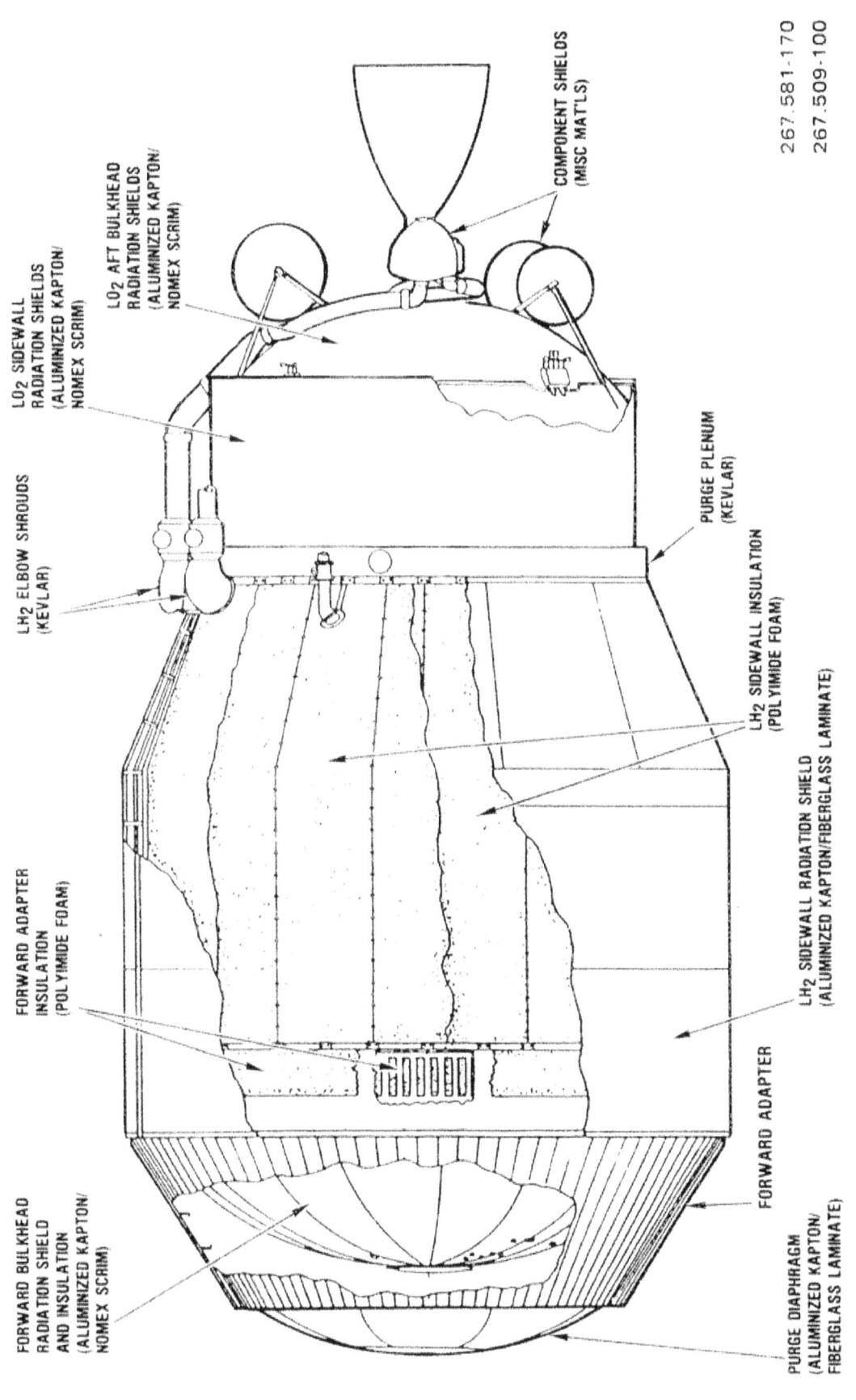

LO2 SIDEWALL
RADIATION SHIELDS
(ALUMINIZED KAPTON/
NOMEX SCRIM)

LO2 AFT BULKHEAD
RADIATION SHIELDS
(ALUMINIZED KAPTON/
NOMEX SCRIM)

COMPONENT SHIELDS
(MISC MAT'LS)

267.581-170
267.509-100

LH2 ELBOW SHROUDS
(KEVLAR)

PURGE PLENUM
(KEVLAR)

LH2 SIDEWALL INSULATION
(POLYIMIDE FOAM)

LH2 SIDEWALL
(POLYIMIDE FOAM)

LH2 SIDEWALL RADIATION SHIELD
(ALUMINIZED KAPTON/FIBERGLASS LAMINATE)

FORWARD ADAPTER
INSULATION
(POLYIMIDE FOAM)

FORWARD ADAPTER

FORWARD BULKHEAD
RADIATION SHIELD
AND INSULATION
(ALUMINIZED KAPTON/
NOMEX SCRIM)

PURGE DIAPHRAGM
(ALUMINIZED KAPTON/
FIBERGLASS LAMINATE)

Figure 3-16. Insulation System for Centaur G-Prime

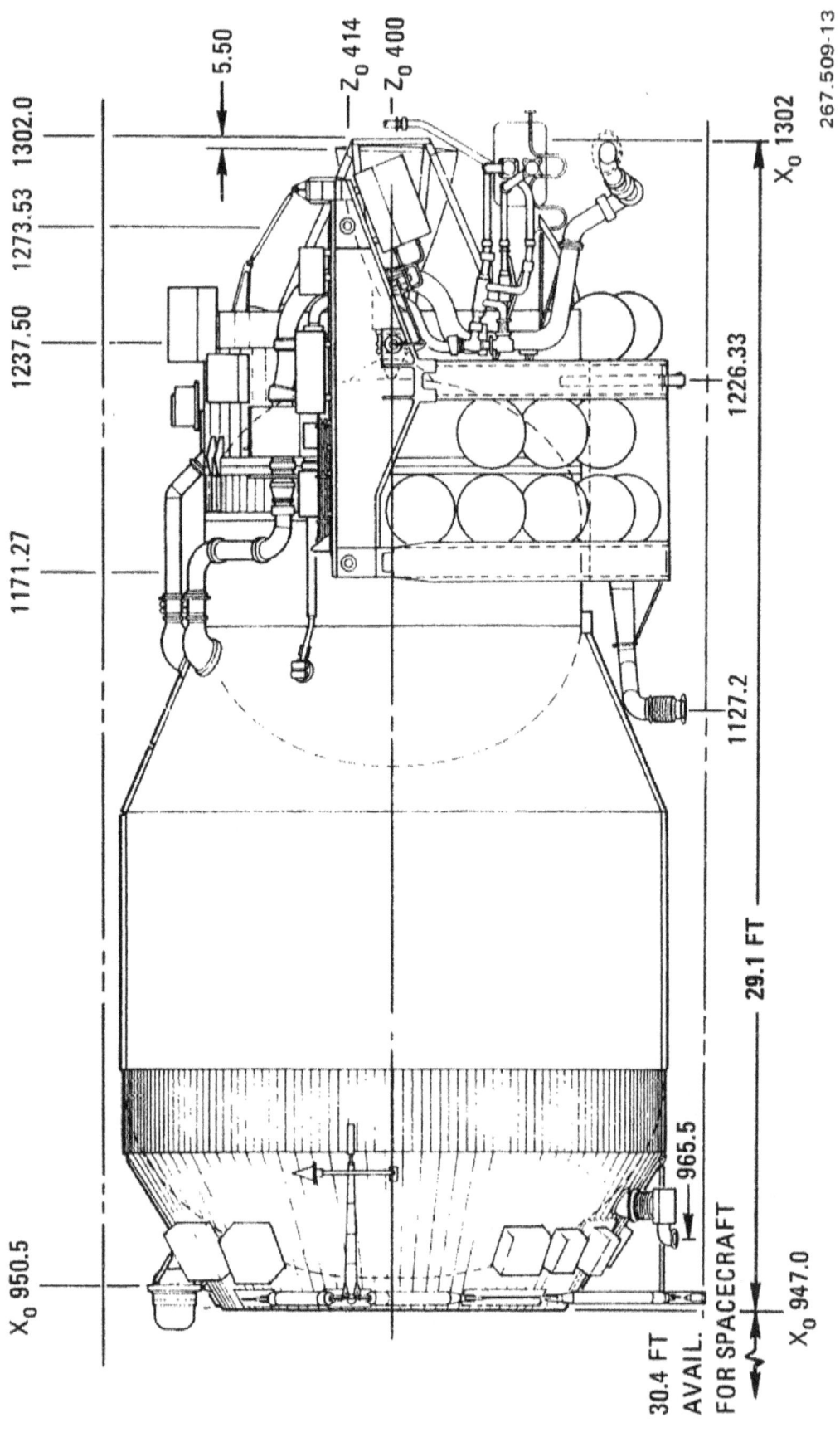

Figure 3-1. Centaur G-Prime Configuration

267.509-13

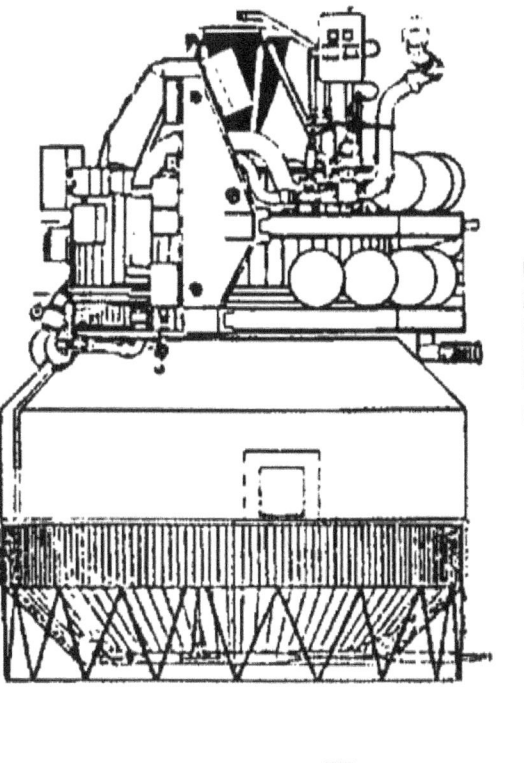

G VEHICLE

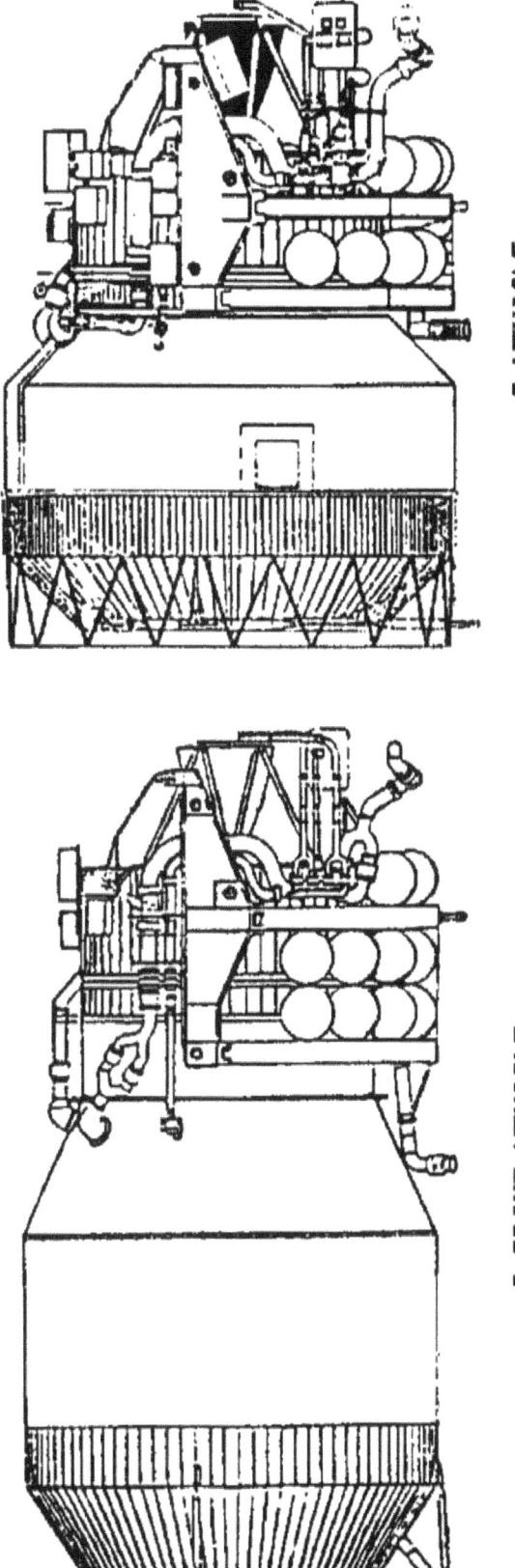

G-PRIME VEHICLE

	G-PRIME VEHICLE	G VEHICLE
LENGTH, FT	30	20
FORWARD ADAPTER CONE, DEG	56.7	46.9
LH$_2$ TANK CONE, DEG	24	45
CENTAUR SUPPORT SYSTEM (CSS) PIN-TO-PIN, IN.	102.26	70.80
CSS He BOTTLES	20	12
FORWARD BULKHEAD MAJOR-TO-MINOR DIAMETER RATIO A/B	1.38	1.58
DRY WEIGHT (VEHICLE), LB	6088	6750
TANKED WEIGHT (VEHICLE), LB	50 270	37 319
CENTAUR INTEGRATED SUPPORT SYSTEM (CISS) WEIGHT, LB	6528	6497

CENTAUR INTEGRATED SUPPORT SYSTEM

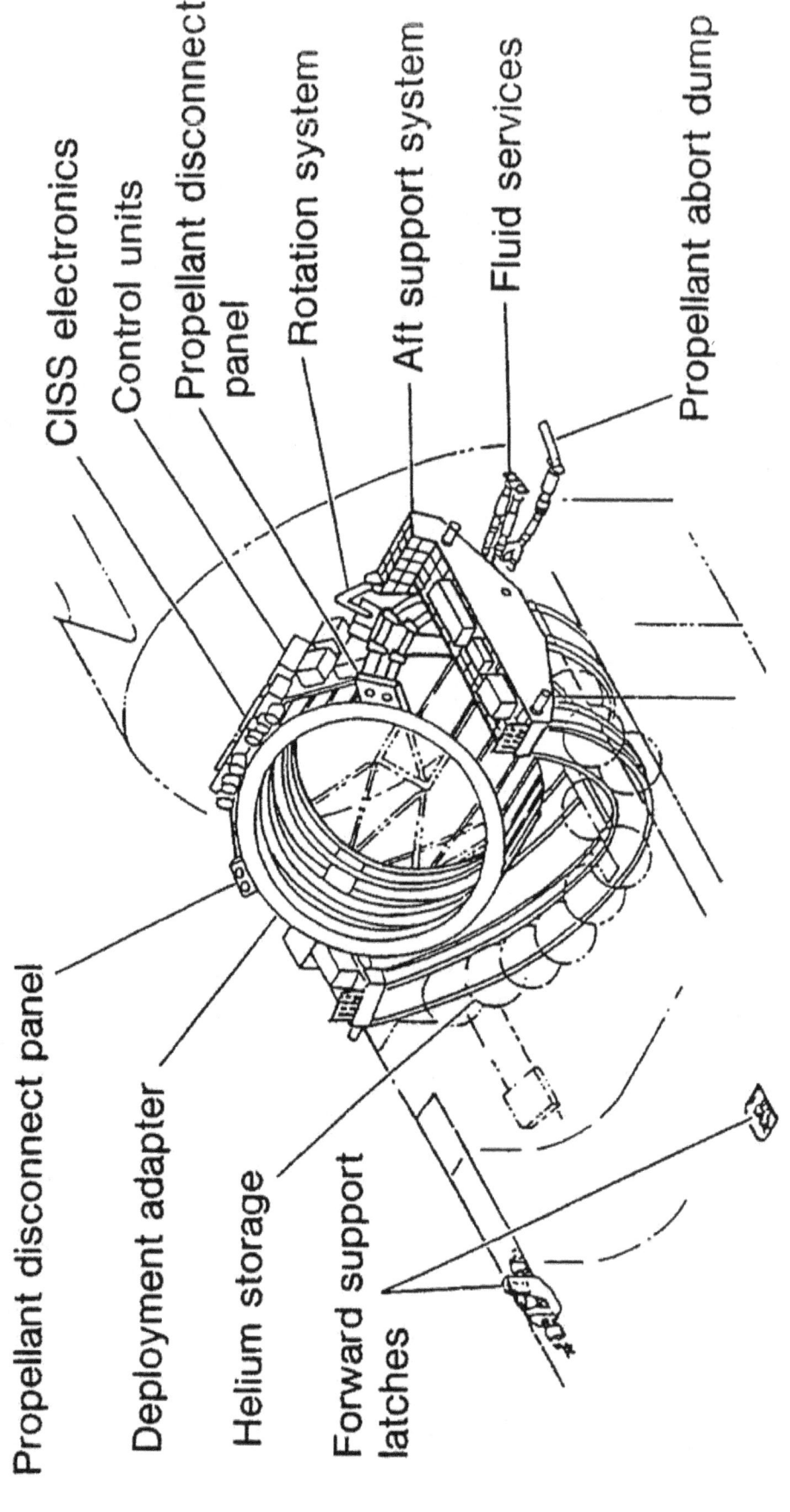

Propellant disconnect panel

Deployment adapter

Helium storage

Forward support latches

CISS electronics

Control units

Propellant disconnect panel

Rotation system

Aft support system

Fluid services

Propellant abort dump

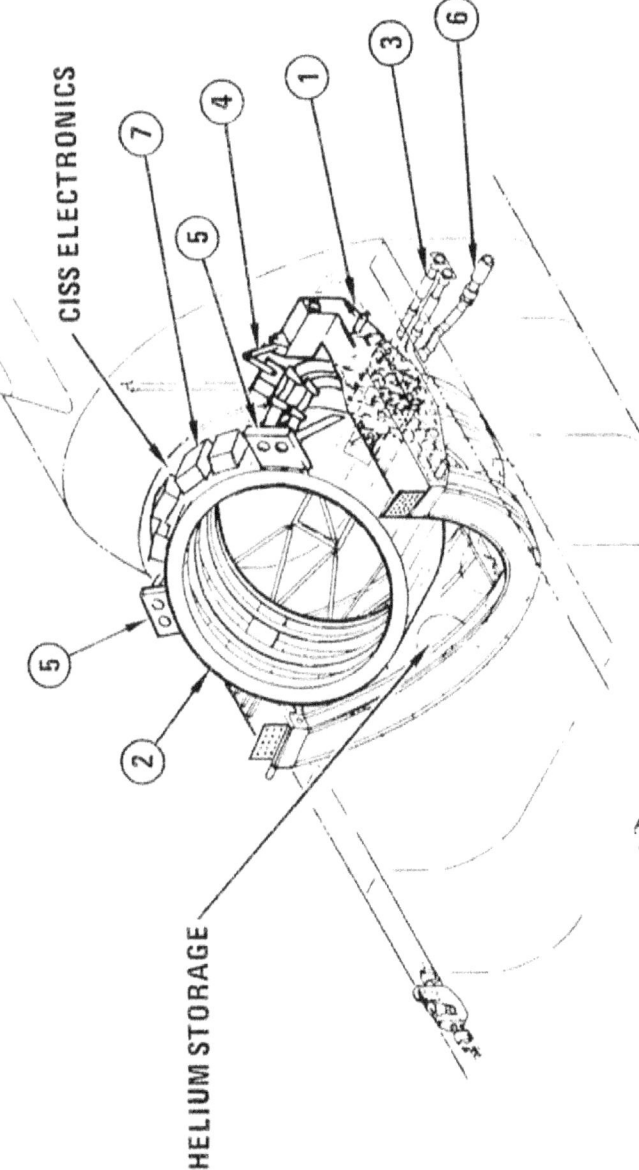

Fig. 3-3. Centaur Integrated Support System (CISS)

267.509-17

CISS ELECTRONICS

HELIUM STORAGE

INTERFACE COMPATIBILITY

1. FIVE-POINT AFT SUPPORT SYSTEM BETWEEN CISS AND ORBITER (STANDARD FITTINGS WITH ADJUSTMENT CAPABILITY ADDED TO FORWARD SILL LATCHES)

2. DEPLOYMENT ADAPTER WITH SPRING THRUST TO EJECT CENTAUR

3. FLUID SERVICING FOR LH$_2$ FILL & DRAIN AND PRELAUNCH TANK VENTING

4. DEPLOYMENT ADAPTER ROTATION SYSTEM (TO 45 DEGREES)

5. FUEL AND OXIDIZER DISCONNECT PANELS

SAFETY CONSIDERATIONS

6. PROPELLANT ABORT DUMP

7. FIVE AUTONOMOUS CONTROL UNITS PROVIDE TWO-FAILURE TOLERANT CONTROL

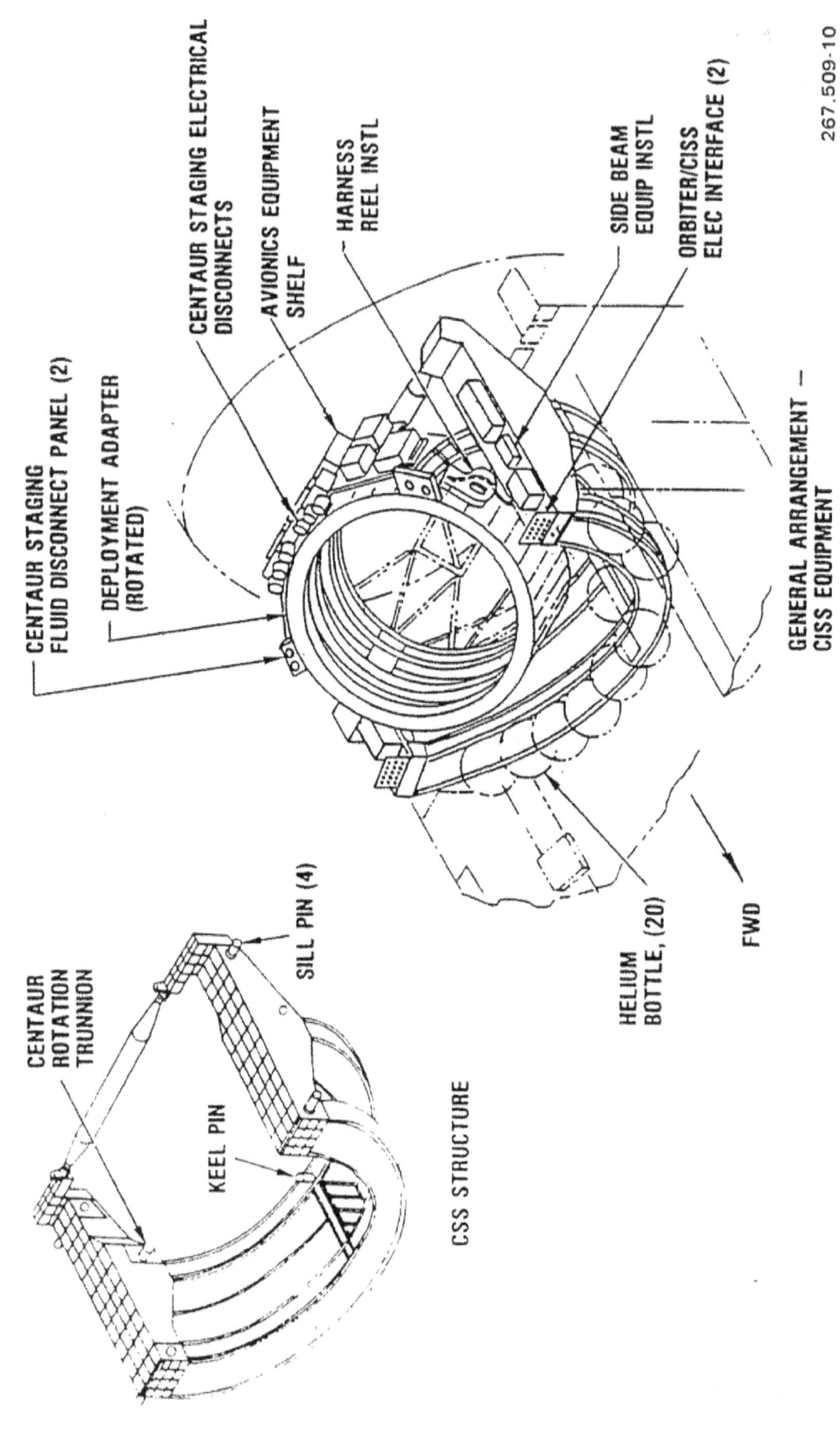

CENTAUR STAGING FLUID DISCONNECT PANEL (2)

DEPLOYMENT ADAPTER (ROTATED)

CENTAUR STAGING ELECTRICAL DISCONNECTS

AVIONICS EQUIPMENT SHELF

HARNESS REEL INSTL

SIDE BEAM EQUIP INSTL

ORBITER/CISS ELEC INTERFACE (2)

GENERAL ARRANGEMENT — CISS EQUIPMENT

HELIUM BOTTLE, (20)

FWD

CENTAUR ROTATION TRUNNION

SILL PIN (4)

KEEL PIN

CSS STRUCTURE

267.509-10

Figure 2-6. Centaur Integrated Support System Design Considerations

9-10

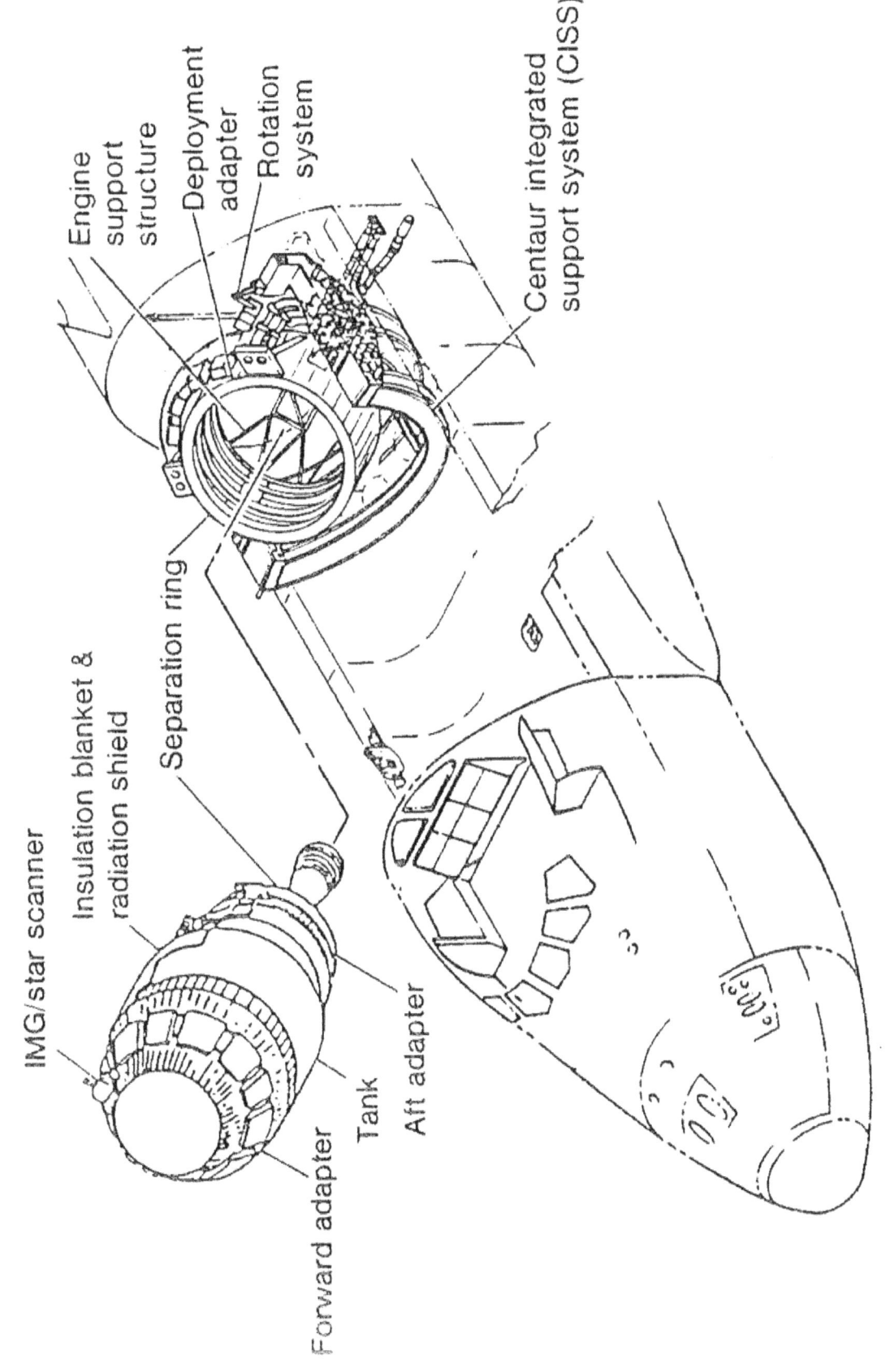

Engine support structure

Deployment adapter

Rotation system

Centaur integrated support system (CISS)

Separation ring

Insulation blanket & radiation shield

IMG/star scanner

Forward adapter

Tank

Aft adapter

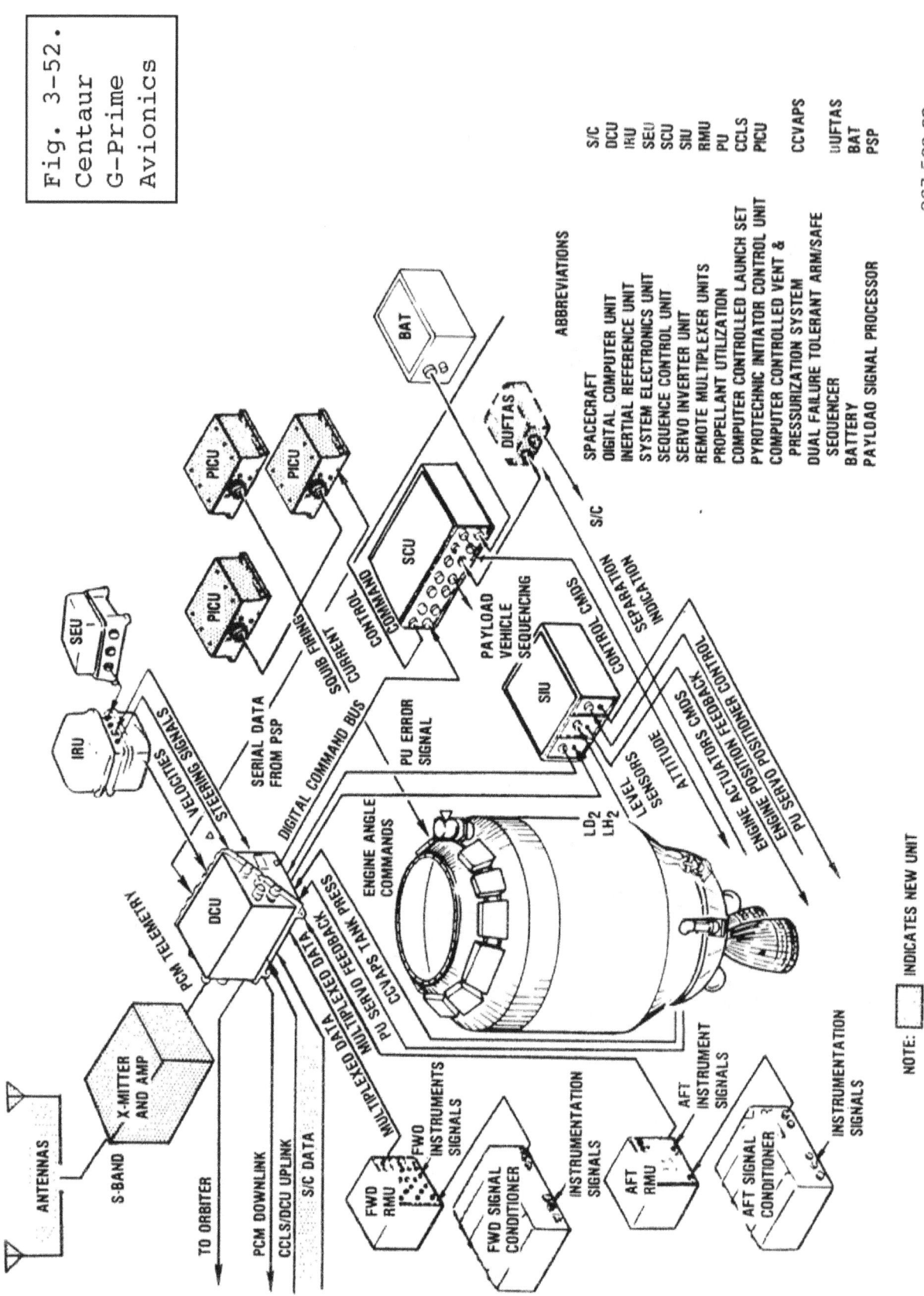

Fig. 3-52.
Centaur
G-Prime
Avionics

267.509-68

ABBREVIATIONS

S/C	SPACECRAFT
DCU	DIGITAL COMPUTER UNIT
IRU	INERTIAL REFERENCE UNIT
SEU	SYSTEM ELECTRONICS UNIT
SCU	SEQUENCE CONTROL UNIT
SIU	SERVO INVERTER UNIT
RMU	REMOTE MULTIPLEXER UNITS
PU	PROPELLANT UTILIZATION
CCLS	COMPUTER CONTROLLED LAUNCH SET
PICU	PYROTECHNIC INITIATOR CONTROL UNIT
CCVAPS	COMPUTER CONTROLLED VENT & PRESSURIZATION SYSTEM
DUFTAS	DUAL FAILURE TOLERANT ARM/SAFE SEQUENCER
BAT	BATTERY
PSP	PAYLOAD SIGNAL PROCESSOR

NOTE: ☐ INDICATES NEW UNIT

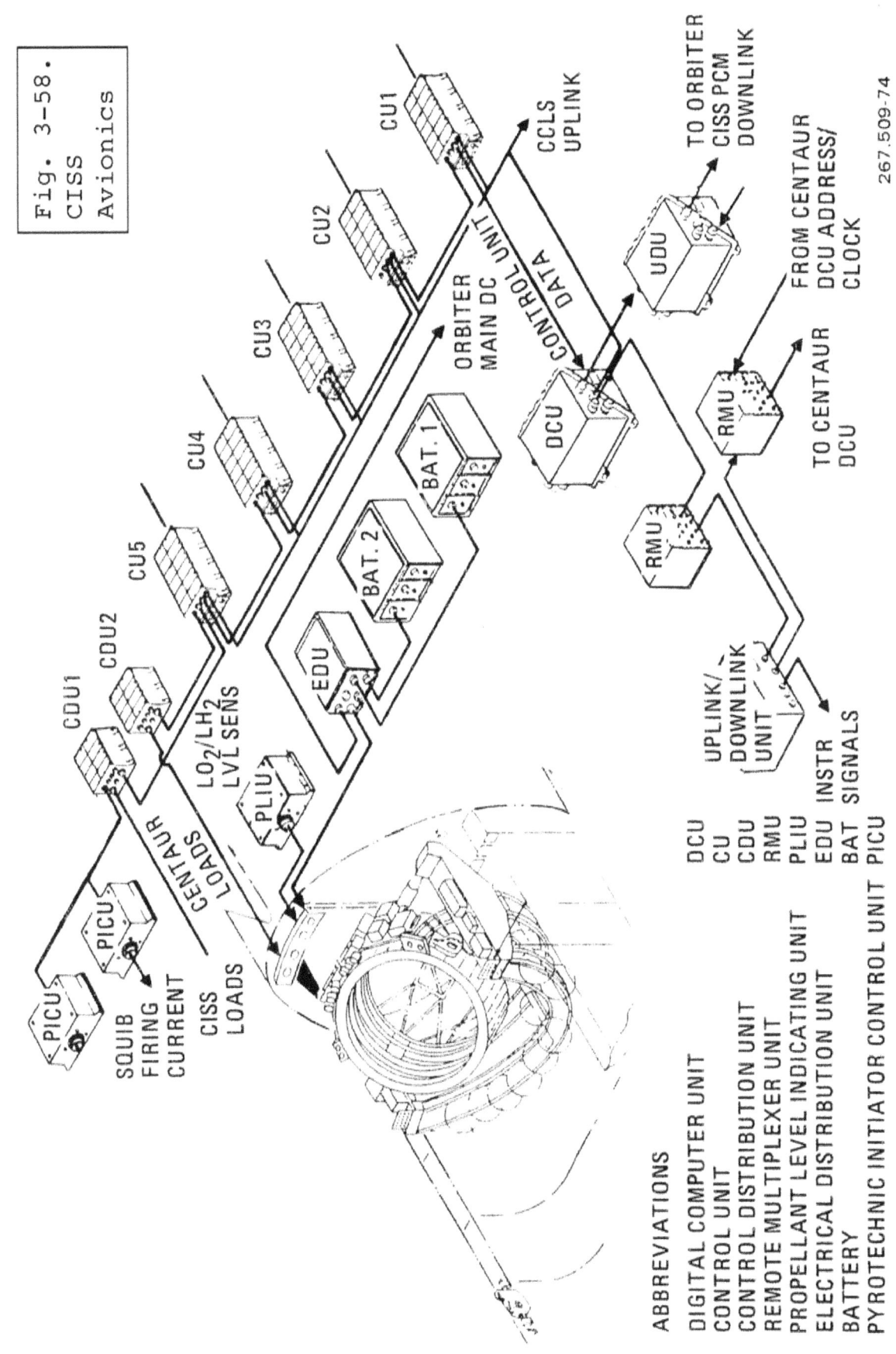

Fig. 3-58.
CISS Avionics

267.509-74

CU1

CU2

CU3

CU4

CU5

CDU1

CDU2

CCLS
UPLINK

ORBITER
MAIN DC UNIT

CONTROL
DATA

UDU

DCU

RMU

RMU

TO ORBITER
CISS PCM
DOWNLINK

FROM CENTAUR
DCU ADDRESS/
CLOCK

TO CENTAUR
DCU

UPLINK/
DOWNLINK
UNIT

LO$_2$/LH$_2$
LVL SENS

BAT. 1

BAT. 2

EDU

PLIU

CENTAUR
LOADS

SQUIB
FIRING
CURRENT

CISS
LOADS

PICU

PICU

PICU

EDU INSTR
BAT SIGNALS

ABBREVIATIONS

DIGITAL COMPUTER UNIT   DCU
CONTROL UNIT   CU
CONTROL DISTRIBUTION UNIT   CDU
REMOTE MULTIPLEXER UNIT   RMU
PROPELLANT LEVEL INDICATING UNIT   PLIU
ELECTRICAL DISTRIBUTION UNIT   EDU
BATTERY   BAT
PYROTECHNIC INITIATOR CONTROL UNIT PICU

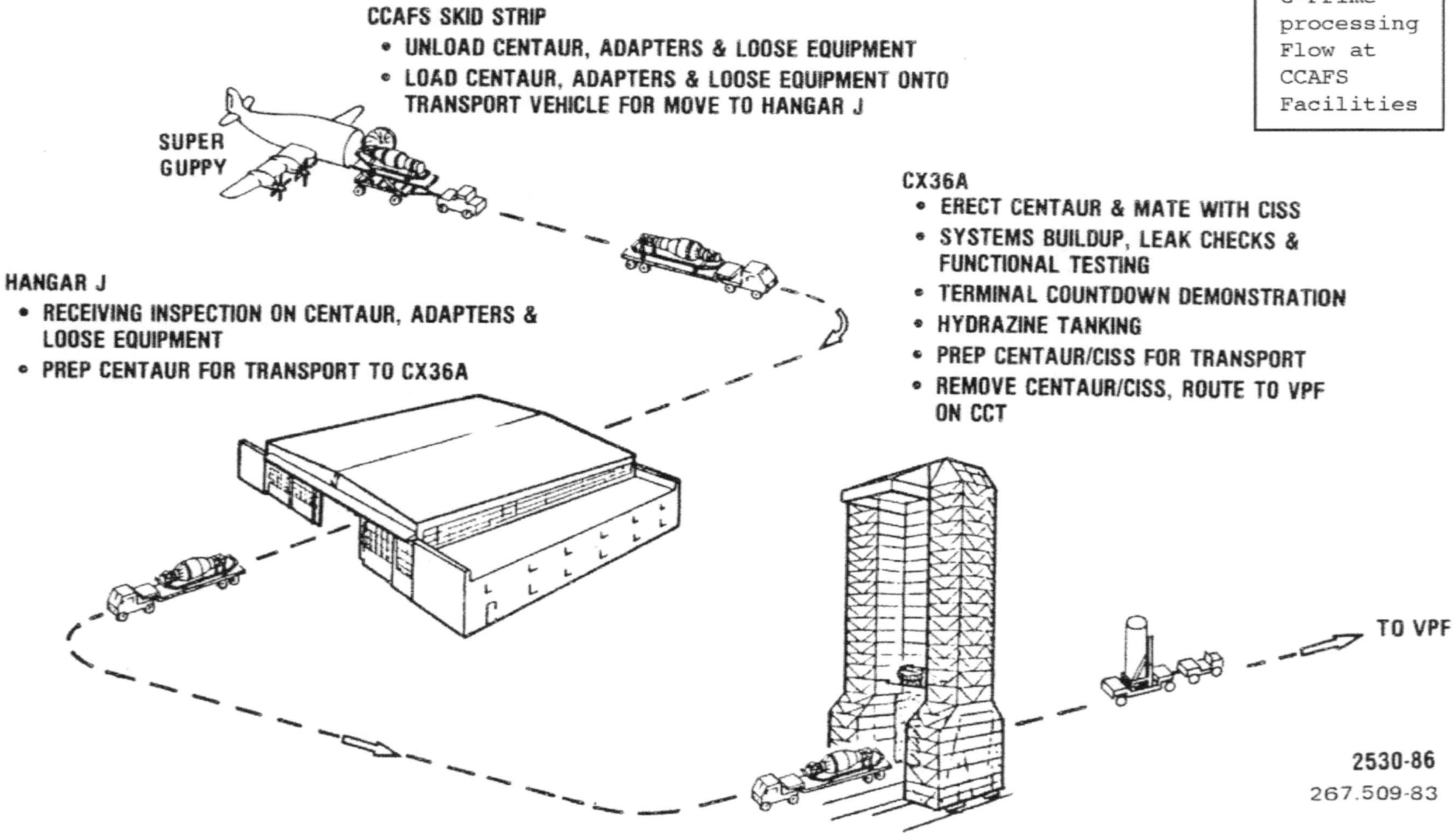

Fig. 7-2. Centaur G-Prime processing Flow at CCAFS Facilities

CCAFS SKID STRIP

- UNLOAD CENTAUR, ADAPTERS & LOOSE EQUIPMENT
- LOAD CENTAUR, ADAPTERS & LOOSE EQUIPMENT ONTO TRANSPORT VEHICLE FOR MOVE TO HANGAR J

SUPER GUPPY

HANGAR J

- RECEIVING INSPECTION ON CENTAUR, ADAPTERS & LOOSE EQUIPMENT
- PREP CENTAUR FOR TRANSPORT TO CX36A

CX36A

- ERECT CENTAUR & MATE WITH CISS
- SYSTEMS BUILDUP, LEAK CHECKS & FUNCTIONAL TESTING
- TERMINAL COUNTDOWN DEMONSTRATION
- HYDRAZINE TANKING
- PREP CENTAUR/CISS FOR TRANSPORT
- REMOVE CENTAUR/CISS, ROUTE TO VPF ON CCT

TO VPF

2530-86
267.509-83

9-14

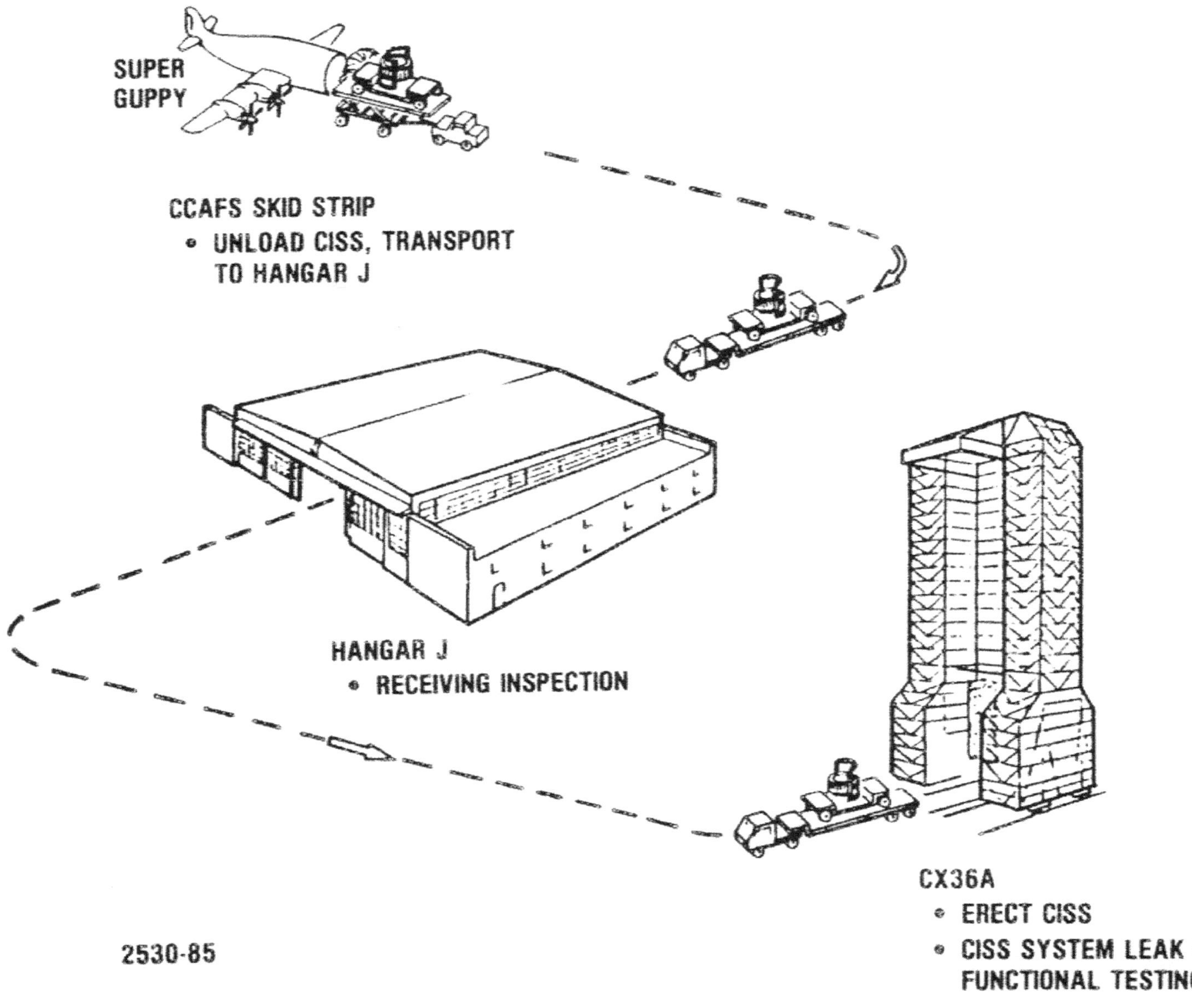

Fig. 7-1. Centaur G-Prime CISS Prelaunch Flowpath

267.509-82

SUPER GUPPY

CCAFS SKID STRIP
• UNLOAD CISS, TRANSPORT TO HANGAR J

HANGAR J
• RECEIVING INSPECTION

CX36A
• ERECT CISS
• CISS SYSTEM LEAK CHECKS & FUNCTIONAL TESTING

2530-85

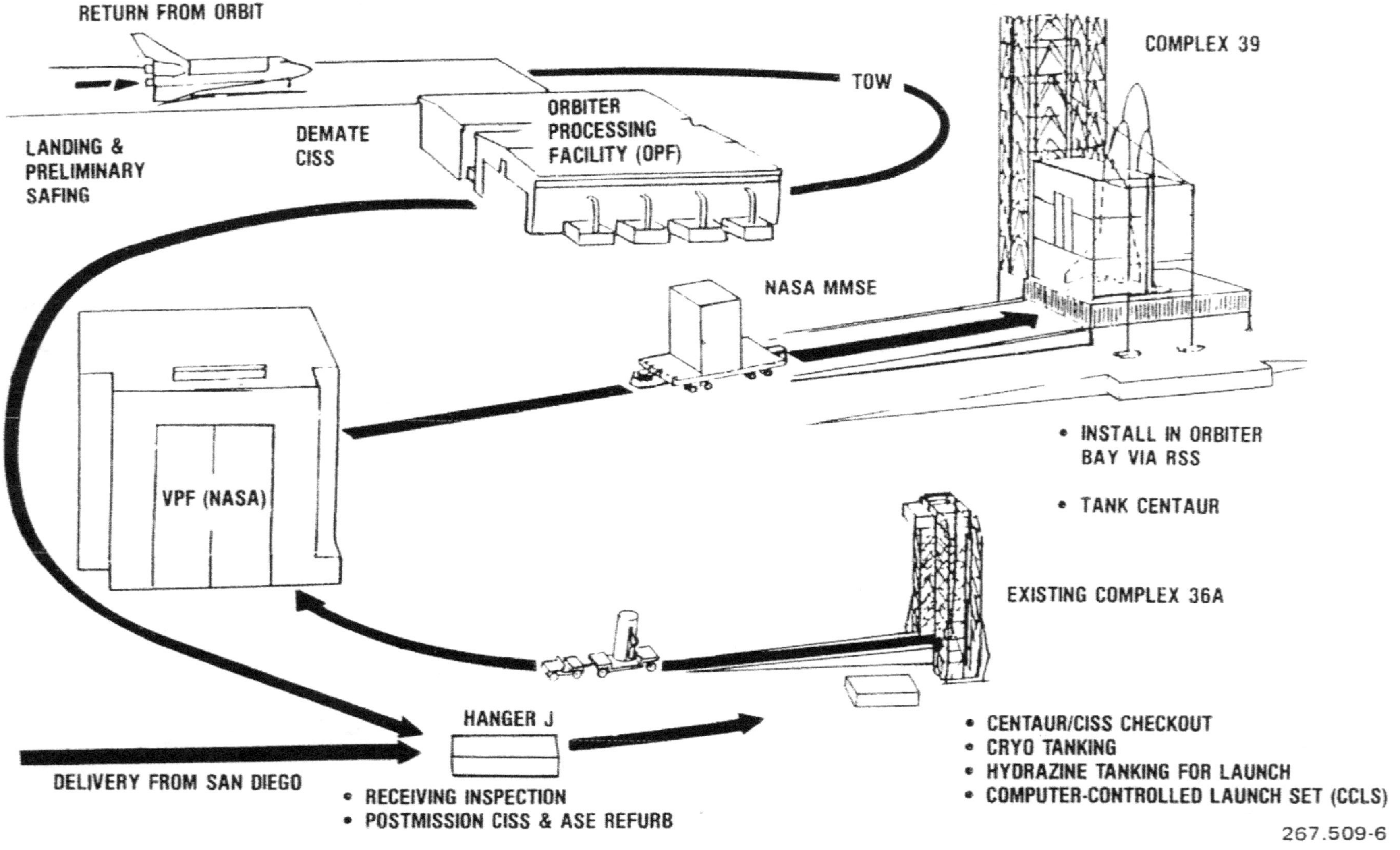

Figure 2-4. Centaur G-Prime Ground Operations at ELS

COMPLEX 39

- INSTALL IN ORBITER BAY VIA RSS
- TANK CENTAUR

EXISTING COMPLEX 36A

- CENTAUR/CISS CHECKOUT
- CRYO TANKING
- HYDRAZINE TANKING FOR LAUNCH
- COMPUTER-CONTROLLED LAUNCH SET (CCLS)

267.509-6

TOW

NASA MMSE

ORBITER PROCESSING FACILITY (OPF)

DEMATE CISS

RETURN FROM ORBIT

LANDING & PRELIMINARY SAFING

VPF (NASA)

HANGER J

DELIVERY FROM SAN DIEGO

- RECEIVING INSPECTION
- POSTMISSION CISS & ASE REFURB

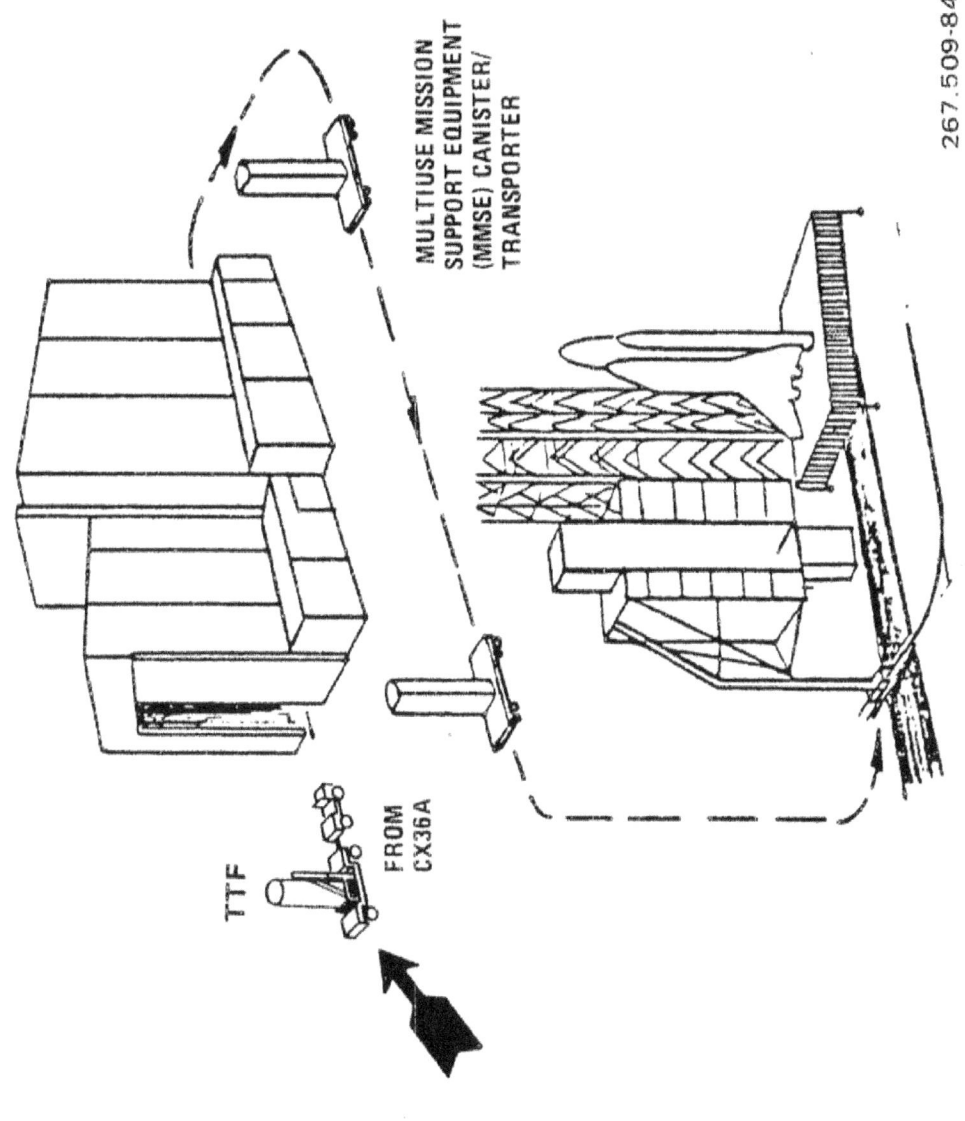

VPF
- RECEIVE CENTAUR/CISS IN TTF
- FINAL CLEAN CENTAUR/CISS
- INSTALL CENTAUR/CISS INTO VPHD
- MATE SPACECRAFT TO CENTAUR
- COMBINED SYSTEMS COMPATIBILITY/
  FUNCTIONAL TESTING
- LOAD CENTAUR/CISS/SPACECRAFT
  (PAYLOAD) INTO CANISTER
- PREP FOR MOVE TO LAUNCH PAD

LAUNCH PAD
- ERECT CANISTER INTO RSS
- REMOVE PAYLOAD FROM CANISTER USING
  THE PGHM
- REMOVE PAYLOAD FROM RSS & INSTALL
  INTO ORBITER CARGO BAY WITH PGHM
- COMBINED SYSTEMS COMPATIBILITY TESTING
- FINAL HAZARDOUS SERVICING & TANKING
- LAUNCH

MULTIUSE MISSION
SUPPORT EQUIPMENT
(MMSE) CANISTER/
TRANSPORTER

TTF

FROM
CX36A

267.509-84

Figure 7-3. Centaur G-Prime Flow Path at KSC

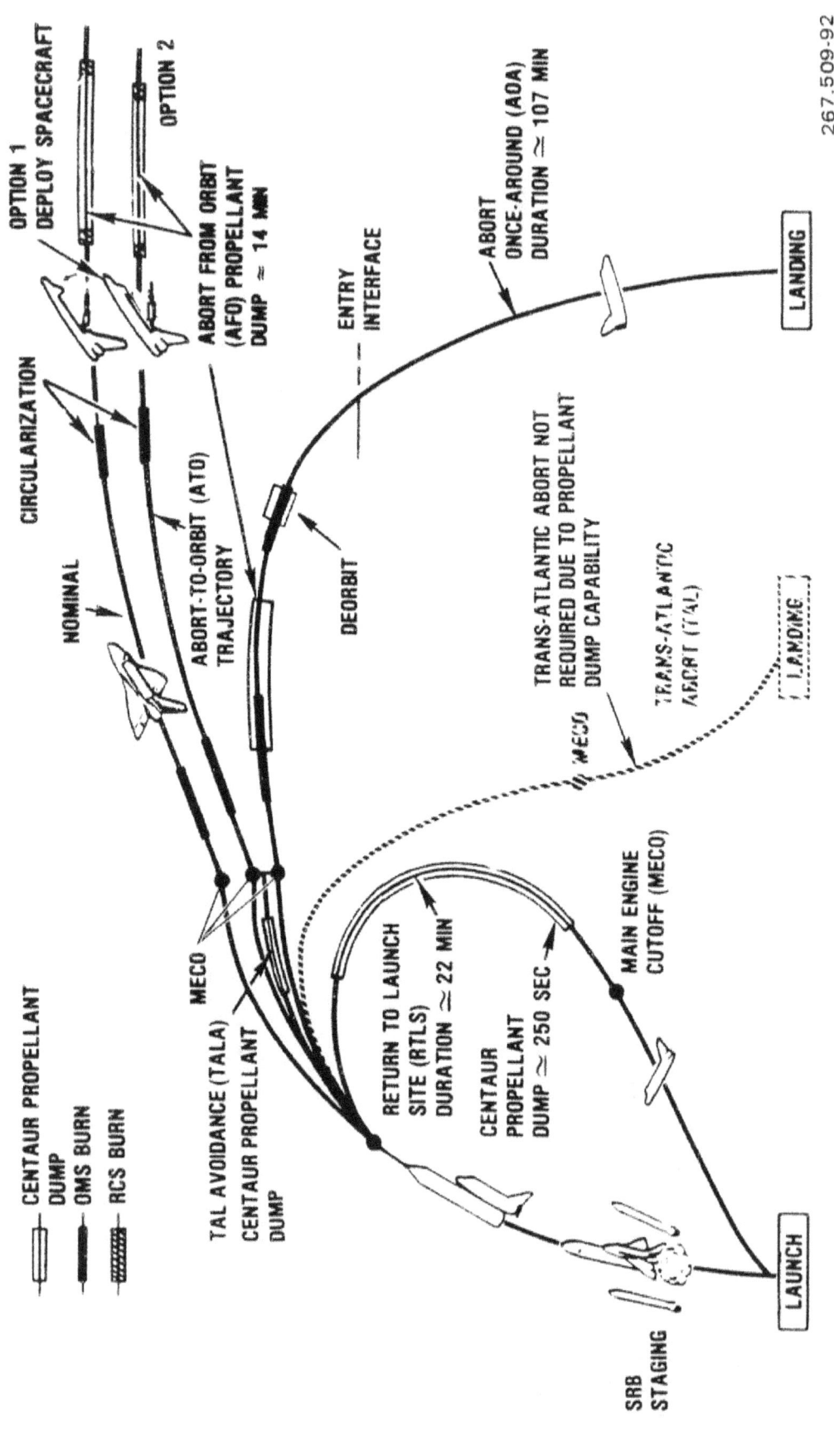

Figure 7-11. Centaur Can Dump Propellants Safely in All Abort Modes

267.509-92

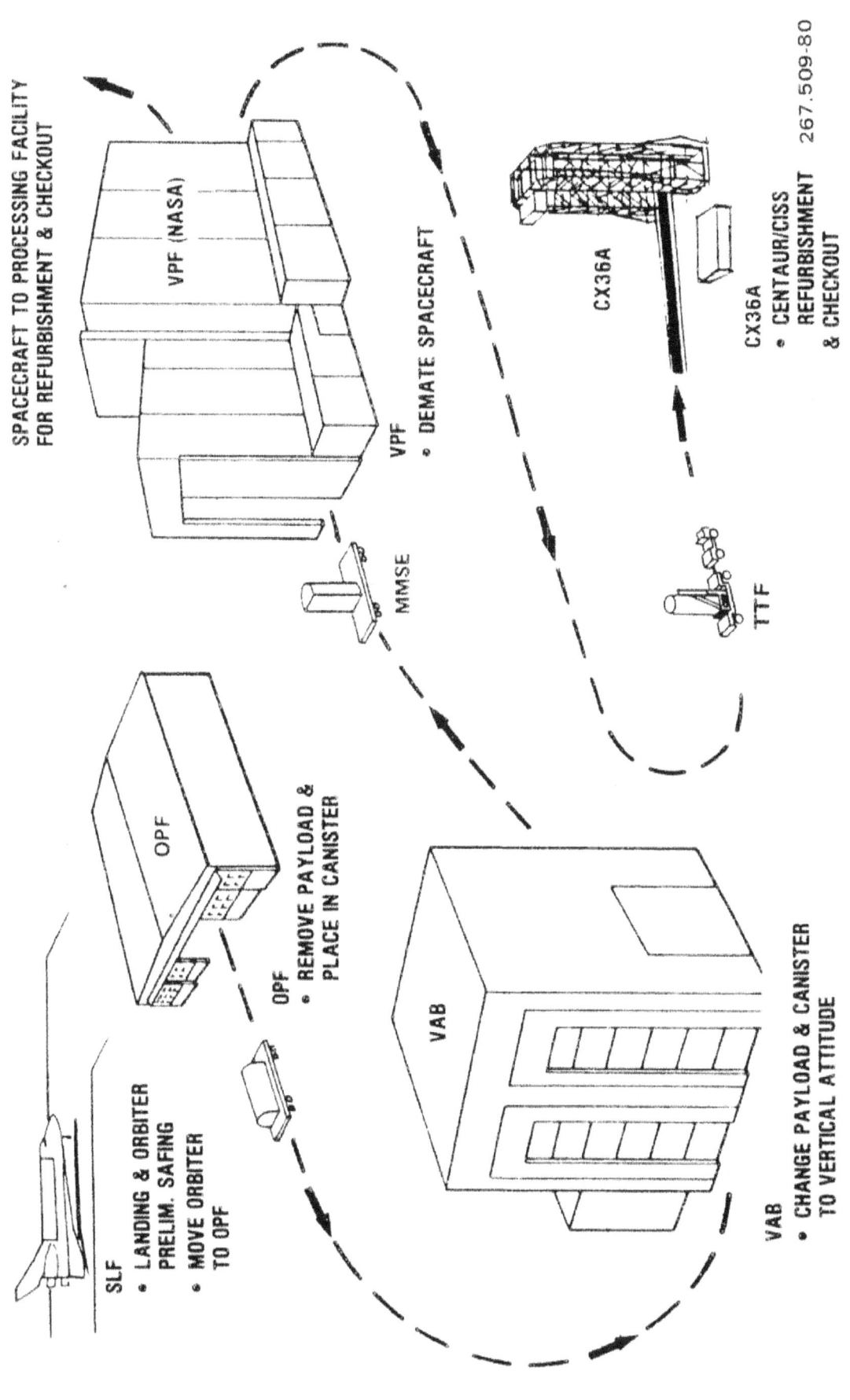

SPACECRAFT TO PROCESSING FACILITY FOR REFURBISHMENT & CHECKOUT

VPF (NASA)

VPF
• DEMATE SPACECRAFT

MMSE

CX36A

CX36A
• CENTAUR/CISS REFURBISHMENT & CHECKOUT

267.509-80

TTF

OPF

OPF
• REMOVE PAYLOAD & PLACE IN CANISTER

VAB

VAB
• CHANGE PAYLOAD & CANISTER TO VERTICAL ATTITUDE

SLF
• LANDING & ORBITER PRELIM. SAFING
• MOVE ORBITER TO OPF

Figure 5-2. Centaur G-Prime Abort Flow Path

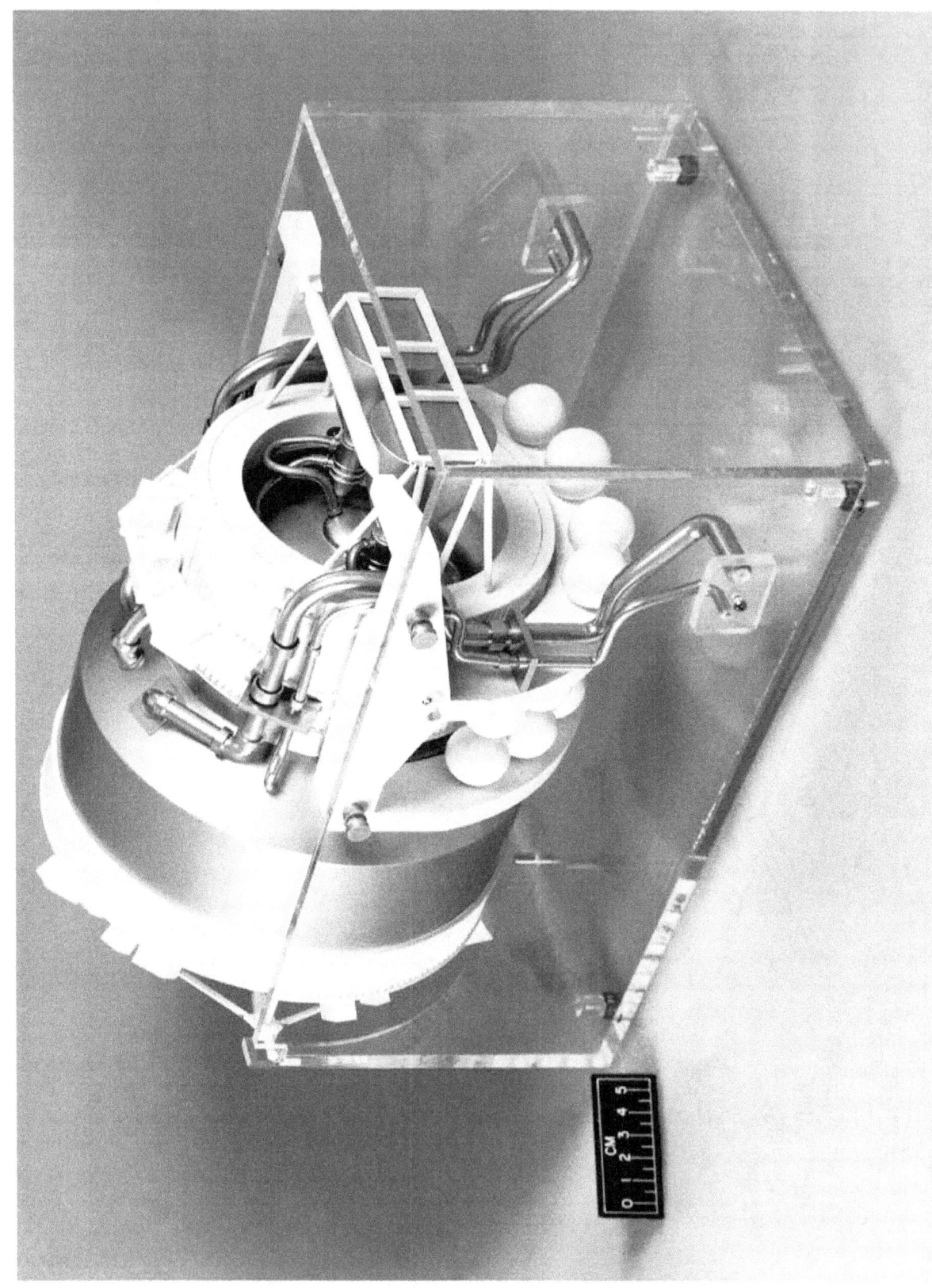

Shuttle-Centaur Model (black and white version of color Image No. C-1985-5360 courtesy NASA).

Two views of a model of the Shuttle/Centaur Test Tower. These images are black and white versions of color Images Nos. C-1984-3331 (left) and C-1984-3330 (right). both courtesy NASA.

Dave Walker (l.) and John Fabian (r.), two of the astronauts scheduled to fly on the first Shuttle-Centaur mission, pose in front of a model of a Centaur stage ready to be deployed from the Space Shuttle. Taken during the rollout ceremony in San Diego at General Dynamics/Convair on August 13, 1985. (Original color Image No C-1985-6216 courtesy NASA; the above black and white version of the original image has been horizontally flipped into its correct orientation, and edited using image processing software, and is a derivative work that is ©2014 Mooncat Collectibles.

Shuttle–Centaur rollout ceremony in San Diego at General Dynamics/
Convair on August 13, 1985. (Original color Image No. C-1985-6212 courtesy NASA.)

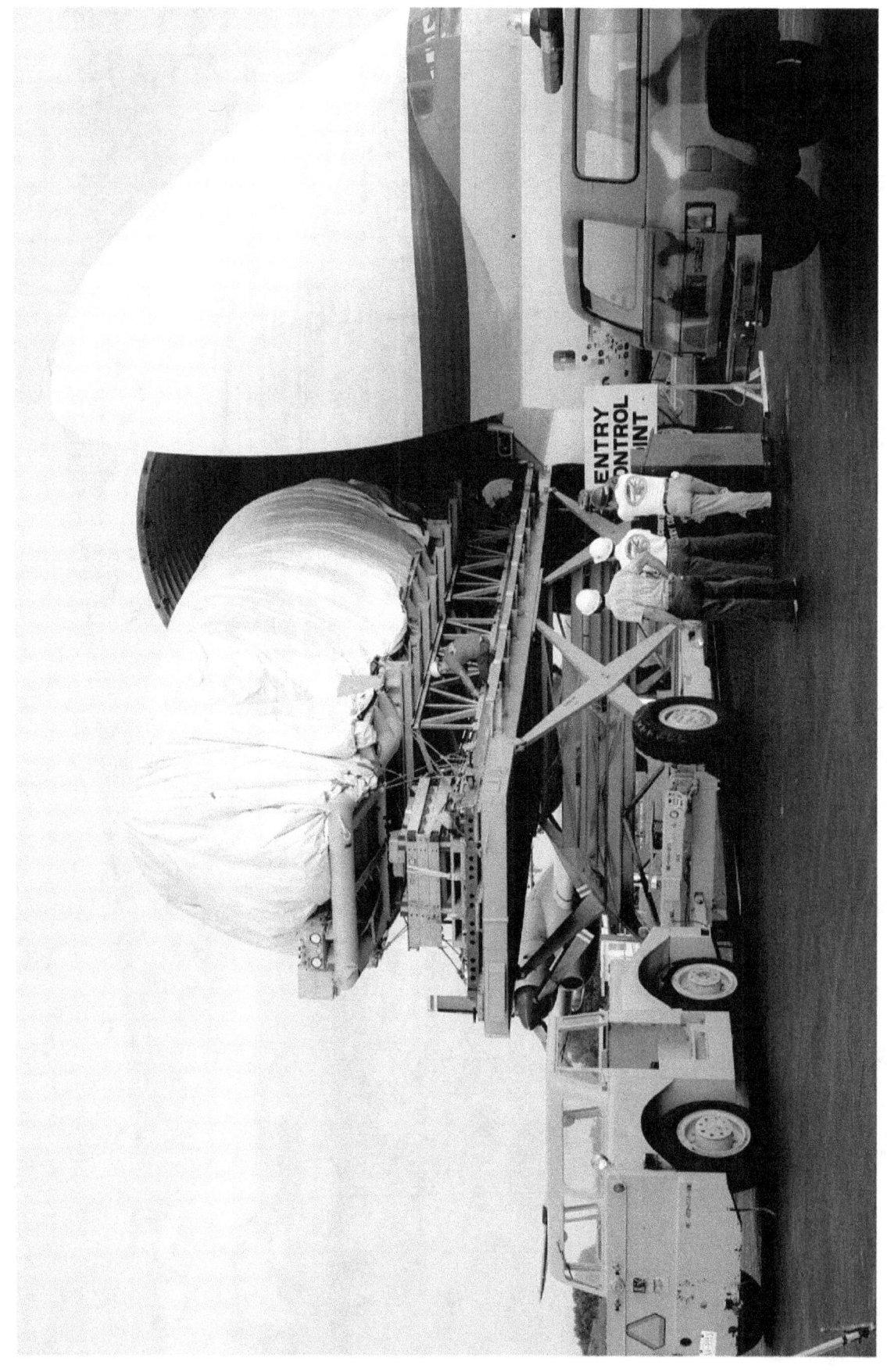

A Shuttle-Centaur stage being unloaded from the Super Guppy at Kennedy Space Centaur. This is a black and white version of color image No. 108-KSC-385C-4352, courtesy NASA.

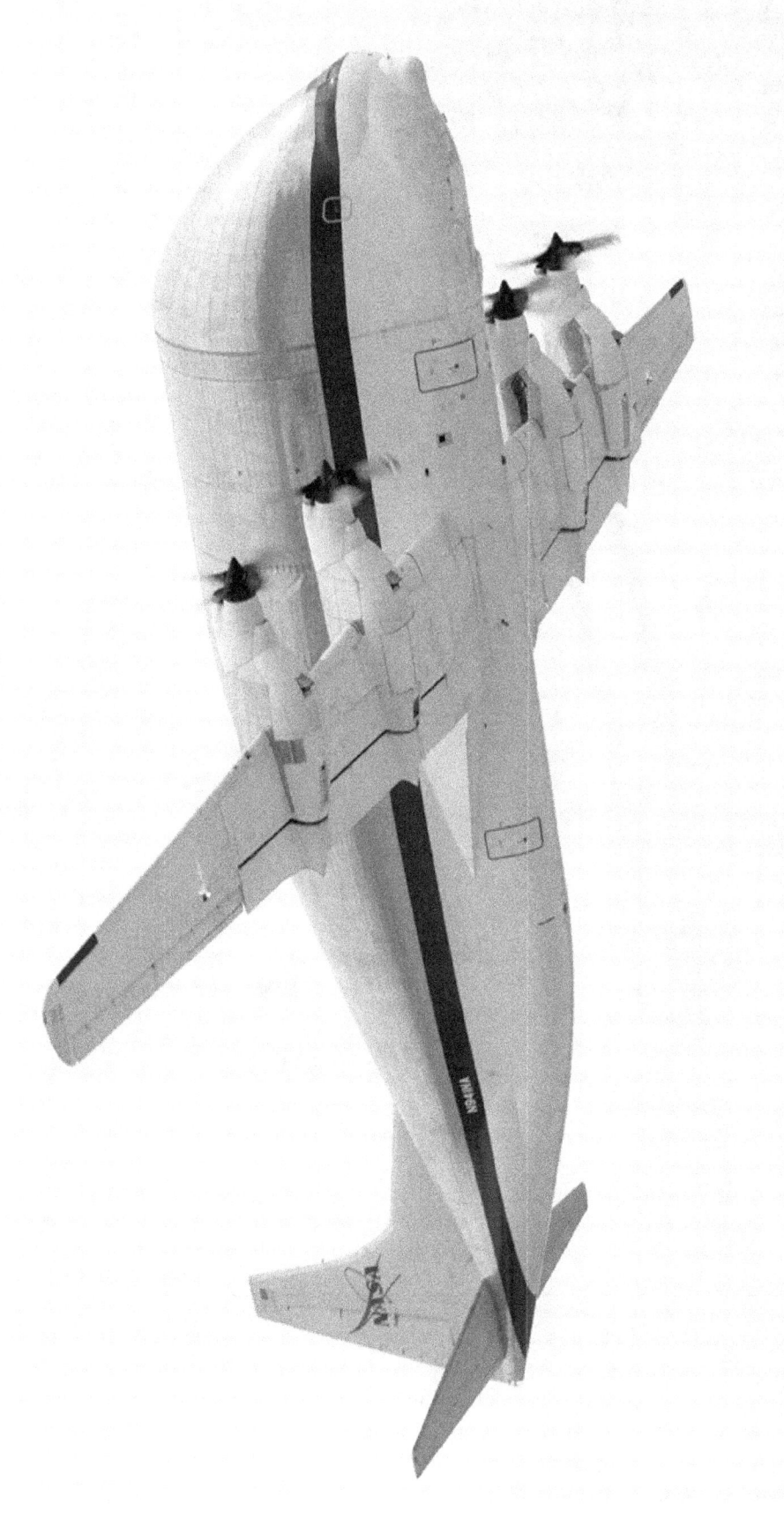

NASA's Super Guppy Turbine cargo plane is used to transport the Shuttle-Centaur stage. This is a black and white version of color image No. EC05-0091-78, courtesy NASA, taken on May 5, 2005.

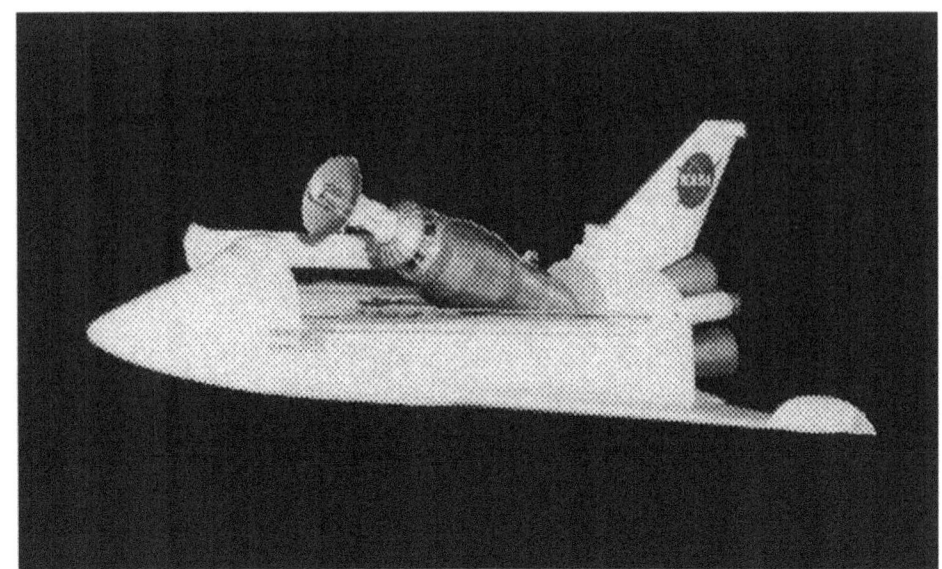

February 22, 1973

Kennedy Space Center, Florida

VOL. 12, No. 4

SPACEPORT NEWS

KSC DIRECTOR DR. KURT H. DEBUS, right, explains the Skylab and Shuttle missions to Senator Frank E. Moss, Chairman of the Senate Committee on Aeronautical and Space Sciences as NASA Administrator Dr. James C. Fletcher, left, looks on. Senator Moss visited the Spaceport February 9 for briefings on NASA and KSC programs and to observe Skylab launch preparations.

Above: An excerpt from the February 22, 1973, issue of *Spaceport News*, the official newsletter of the Kennedy Space Center. The lower picture shows Drs. Debus and Fletcher of NASA explaining models of the two post-Apollo space projects, Skylab and the Space Shuttle, to a Senator. The model of the Space Shuttle, enlarged in the upper panel, includes an upper stage that has the morphology of the Shuttle-Centaur stage. Note that this early version of the Space Shuttle includes winglets. Original document courtesy NASA; the upper panel is a derivative work that is ©2014 Mooncat Collectibles.

NASA ──────
Lyndon B. Johnson Space Center

Space News Roundup

Vol. 20 No. 12   June 19, 1981   National Aeronautics and Space Administration

Centaur Studied as Shuttle Stage

NASA's Lewis Research Center, Cleveland, has awarded four letter contracts totaling $7,483,000 for design and development of a modified Centaur launch vehicle and related components for use as an upper stage in the Space Shuttle.

Under a $1,545,000 contract with Teledyne Industries Inc., Northridge, Calif., five digital computer units and nine remote multiplexer units will be designed and developed. The digital computer is the on-board computer which the Centaur vehicle utilizer during flight for control and operation without ground commands. The remote multiplexer units comprise the basic airborne data information system to supply in-flight data during launch. Work will be performed at the contractor's plant in Northridge.

Under a $1,593,000 contract with Honeywell Inc., Avionics Division, St. Petersburg, Fla., three inertial measurement groups will be designed and developed. These are part of the self-contained automatic navigation and guidance system. Work will be performed at St. Petersburg.

Under a,$3,412,000 contract with General Dynamics Corporation, Convair Division, San Diego, two modified Centaur vehicles will be designed and developed, with work to be performed at San Diego.

All work under these contracts was scheduled to begin June 1 and continue through Sept. 30.

Under a $933,000 contract with United Technologies Corporation, Pratt & Whitney Aircraft Group, West Palm Beach, Fla., four RL10A-3-3A rocket engines will be built. Primary thrust for the Centaur is provided by two of these engines which develop 33,000 pounts total thrust. The engines are regeneratively cooled and turbopump fed. Work will be done at West Palm Beach and will begin on August 1 and will continue through Sept. 30.

All of these contracts are in support of the Galileo mission to Jupiter scheduled for launch in 1985 and the International Solar Polar Mission in 1986.

For use with these missions, the Centaur will be an adaptation of the vehicle that has flown as an upper stage for both the Atlas and the Titan boosters over the last 15 years on the Mariner missions to Mars and Venus, the Pioneer missions to Jupiter and Venus, the Viking and Voyager missions, and the cooperative Helios mission with West Germany. Centaur has also flown NASA low earth orbit missions.

Above: An excerpt from the June 19, 1981, issue of Space News Roundup, the newsletter of NASA's Johnson Space Center, announcing the contracts for the first hardware for Shuttle-Centaur. Courtesy NASA.

CENTAUR FOR THE 1980s

JOHN E. NIESLEY

Advanced Systems Project Engineer
Advanced Centaur Programs
General Dynamics Convair Division
San Diego, California

ABSTRACT

Centaur is currently the world's only operational high-energy upper stage, and is the United States primary upper stage for launching solar system probes, large geosynchronous communication satellites, and observatories to study the farthest limits of space. Centaur is currently launched on Atlas, but has also flown with the larger Titan booster. NASA recently decided to integrate Centaur with the Space Shuttle for future solar system exploration missions.

Current status of the Centaur program is discussed including: vehicle characteristics, planned performance improvements, and launch schedules. Modifications required to integrate Centaur with Shuttle and the resulting capabilities are discussed.

INTRODUCTION

Centaur development began in 1958 when General Dynamics/Convair was awarded a contract to develop the first space vehicle to use liquid hydrogen fuel. Because NASA's Lewis Research Center (LeRC) did much of the pioneering work in liquid hydrogen technology, LeRC was later assigned technical management of Centaur and contributed to the first successful launch in 1963. This successful working relationship continues today. After completing the development phase in 1966, the resulting operational vehicle, called Centaur D, was launched 21 times on Atlas.

In the early 1970s, Centaur electronics and guidance systems were completely modernized. A new high-speed digital computer was added that permits extensive use of software to perform functions previously requiring hardware, thus simplifying new mission adaptation. Computer controlled launch set (CCLS) was added to provide rapid automatic checkout of the Centaur and diagnostic capabilities for anomalies. This new version of Centaur, designated D-1, was integrated with both the Atlas vehicle and the more powerful Titan booster, and has flown 32 operational missions.

Currently, Centaur is undergoing additional performance improvements for Intelsat, which will enhance its capabilities for the 1980s. NASA has also recently decided to integrate Centaur with the Space Shuttle for solar exploration missions, large future geosynchronous commercial satellites, and potential Department of Defense (DoD) missions. These applications will ensure continued use of Centaur through the remainder of the 1980s.

CENTAUR RECORD

A little over two decades ago, Centaur was conceived as the upper stage for United States' solar system exploration and geosynchronous communications satellites. Today that dream has truly been fulfilled by the accomplishments of this vehicle. Centaur has launched 22 solar system exploration missions including Voyager, Viking, Helios, Mariner, Surveyor and Pioneer. Its selection by NASA for launching the Galileo and International Solar Polar missions from Space Shuttle means continuation of this enviable record. In addition, 24 geosynchronous communication satellites have been launched as well as 6 space observatories (Figure 1). Centaur has flown 55 times with Atlas and 7 times on Titan for a total of 62 flights and is now ready for integration with the Space Shuttle.

During the past fifteen years of operational flights, Centaur has established itself as a reliable upper stage. 96% of all operational flights were successful, with 100% or 36 consecutive successes since 1971. The Pratt and Whitney RL-10 engines have a perfect flight success record and the current Centaur D-1 guidance and navigation system has also performed 100% successfully on all countdowns and launches as indicated in Figure 2.

ATLAS CHARACTERISTICS

The Atlas vehicle that boosts Centaur is a stage-and-a-half configuration in which all engines are ignited on

Missions

- Solar system exploration (22)
  - Voyager, Viking, Helios
  - Mariner, Pioneer, Surveyor
- Communications(25)
  - Intelsat, Fltsatcom, Comstar
- Astronomy (6)
  - HEAO, OAO

Launch platforms

- Atlas
- Titan

Figure 1. Centaur enabled the United States to achieve many dramatic firsts in space and provided a valuable capability for geosynchronous missions.

Vehicle operational successes

- 96% overall
- 100% since 1971
  (36 consecutive successes)

P&W RL-10 engine

- 100% successful flight record
  (66,000 sec in space)

Guidance & Navigation — D-1
(Honeywell IRU & Teledyne DCU)
- 100% flight & countdown success
  (33 missions with 450 operational hours)

Figure 2. Centaur's success record proves its reliabilty as a high-energy upper stage.

the ground and share common propellant tanks. The booster engines are jettisoned at approximately 140 seconds into the flight when vehicle acceleration reaches 5.5 g. The sustainer engine continues to burn until propellant depletion occurs about 110 seconds later. Two small vernier engines burn throughout the booster phase and provide all roll control during the sustainer phase. All engines use liquid oxygen (LO2) and RP-1 fuel, which is similar to kerosene. Vehicle characteristics are shown in Figure 3.

The Atlas vehicle is 10 ft in diameter and approximately 70 ft in length, not including the interstage adapter. Tanks are made of thin-walled stainless steel bands which are welded together and pressure stabilized for structural strength. A helium pressurization system maintains tank pressure for structural integrity and turbopump pressure head during flight. Vehicle

control is accomplished by gimbaling the Atlas engines during flight under direction of the Centaur guidance and navigation system.

CENTAUR CHARACTERISTICS

Centaur is a high energy upper stage powered by two Pratt & Whitney RL-10 engines developing 33,000 lb total vacuum thrust at a rated Isp of 446 seconds (see Figure 4). The stage burns 30,000 lb of liquid hydrogen (LH2) and liquid oxygen (LO2) propellants. Tanks are made of thin-walled type 301 stainless steel welded construction that is pressure stabilized. They are separated by a double-wall vacuum-insulated intermediate common bulkhead and pressurized with gaseous helium. Until now, tank-mounted boost pumps have been used to provide the required engine inlet pressures. The boost pumps are driven by turbines

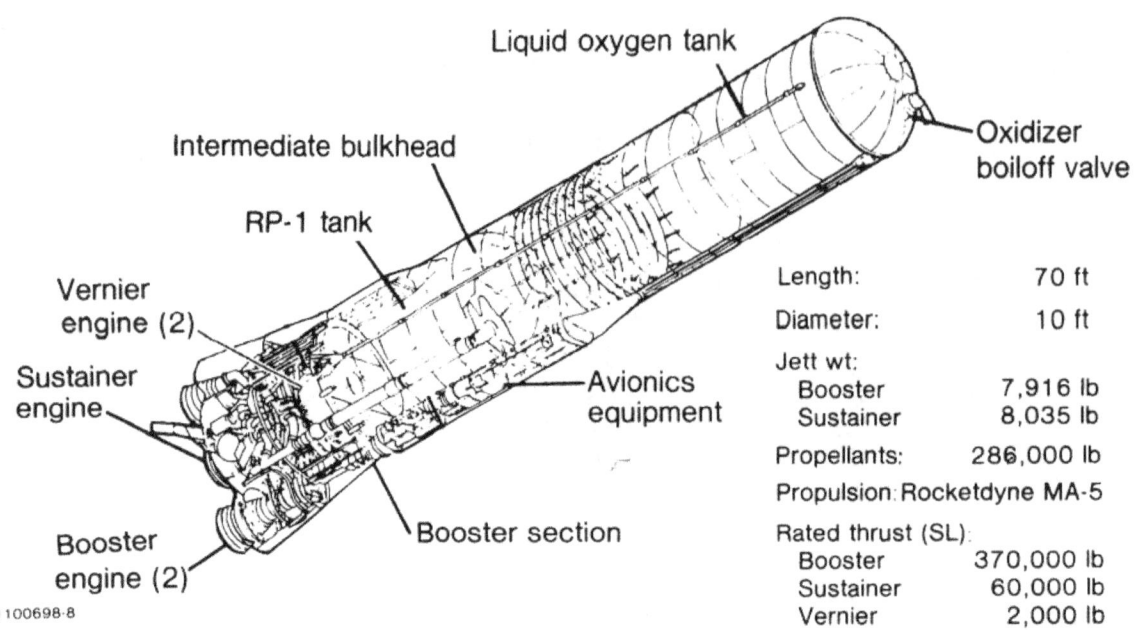

Length:	70 ft
Diameter:	10 ft
Jett wt:	
Booster	7,916 lb
Sustainer	8,035 lb
Propellants:	286,000 lb
Propulsion:	Rocketdyne MA-5
Rated thrust (SL):	
Booster	370,000 lb
Sustainer	60,000 lb
Vernier	2,000 lb

)1100698-8

Figure 3. The Atlas booster vehicle is a state-and-a-half configuration with a sustainer engine that continues to burn 90 seconds after the initial booster engines are jettisoned.

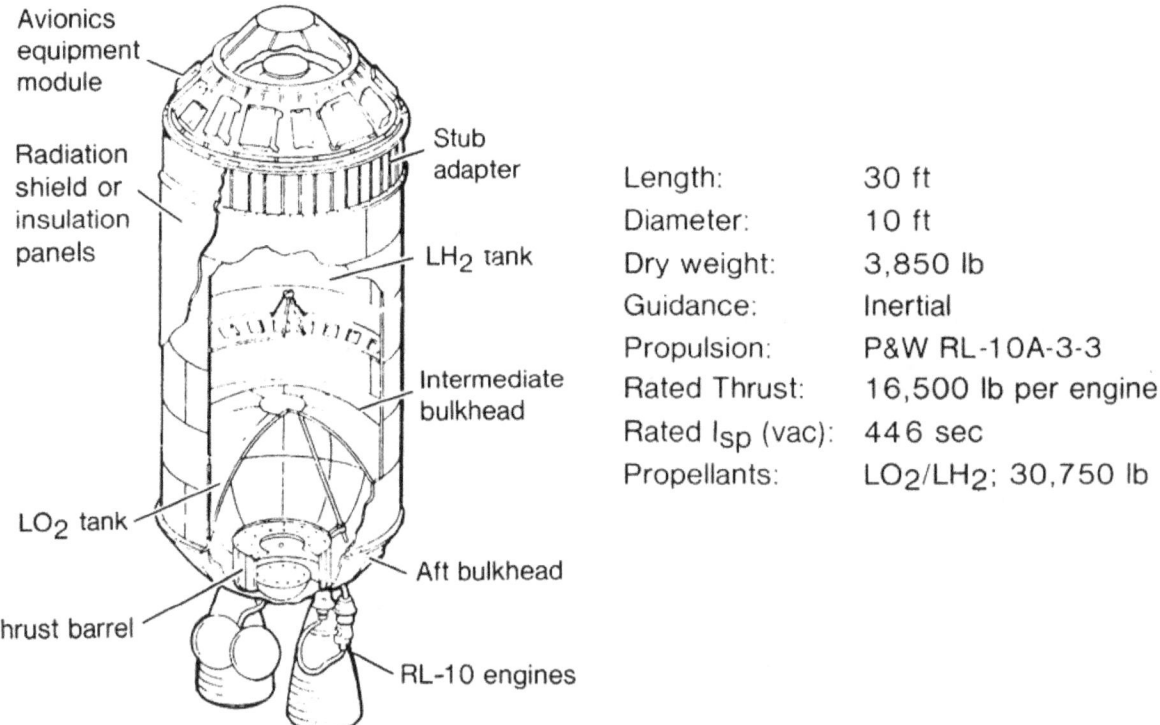

Length:	30 ft
Diameter:	10 ft
Dry weight:	3,850 lb
Guidance:	Inertial
Propulsion:	P&W RL-10A-3-3
Rated Thrust:	16,500 lb per engine
Rated I_{sp} (vac):	446 sec
Propellants:	LO_2/LH_2; 30,750 lb

Figure 4. Centaur characteristics.

powered by hydrogen peroxide, the same monopropellant used in the reaction control system. Beginning with vehicle AC-62, minor tank and engine changes have allowed the elimination of boost pumps. Reduced cost and improved reliability and performance will result. Associated changes are gaseous hydrogen engine bleed for LH_2 tank pressurization during engine operation, and hydrazine monopropellant for the reaction control system.

The Centaur integrated astrionics system is illustrated in Figure 5. The heart of this system is the Teledyne Digital Computer Unit (DCU) which has 16,000

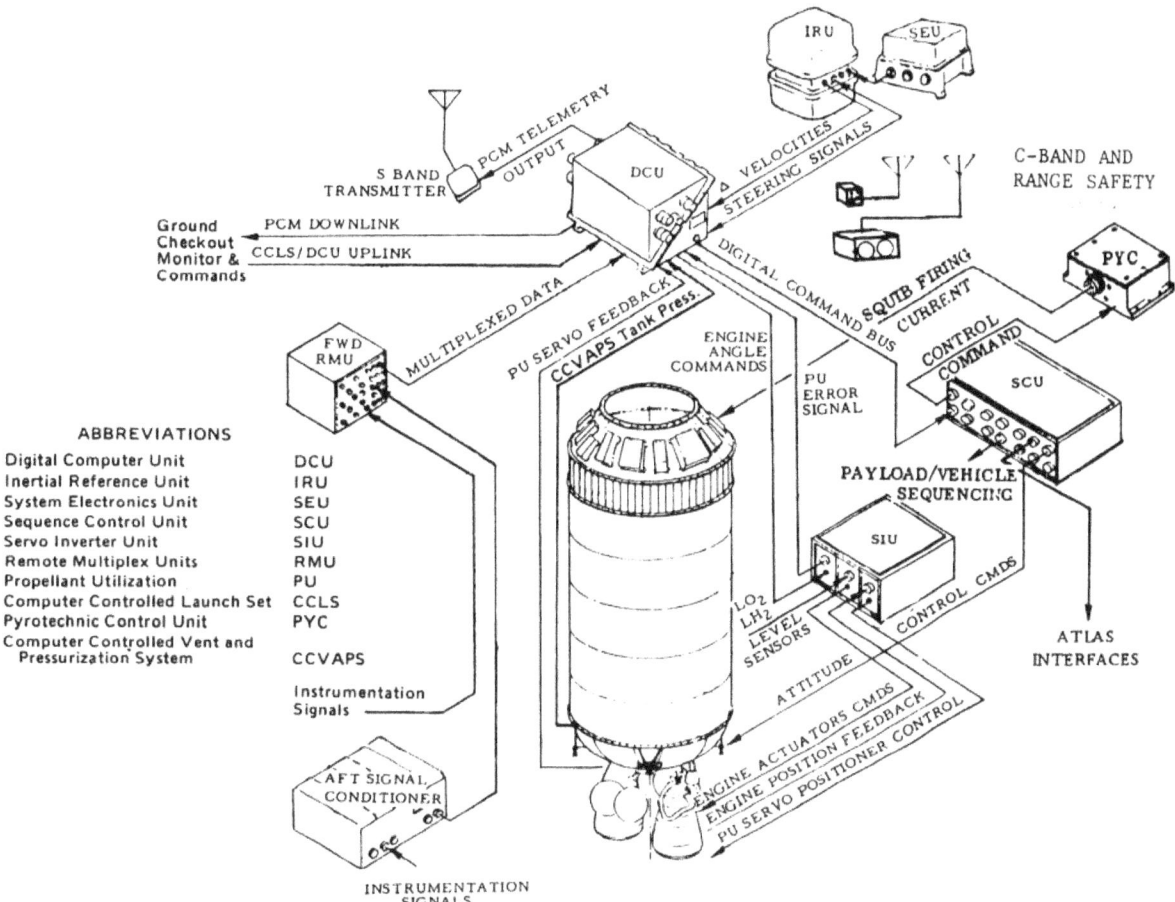

Figure 5. Centaur astrionics system.

words of memory, fast execution speed, and extensive input-output capabilities allowing it to perform many functions that previously required separate hardware.

A Honeywell Inertial Reference Unit (IRU) containing four-gimbals, three reference gyros, and three pulse-balanced accelerometers provides the required navigation data for the DCU. The DCU generates output commands to provide vehicle stability and guidance steering that are routed through the servo inverter unit (SIU) during powered flight and through the sequence control unit (SCU) during coast phase. Centaur flight software is modularized into several special-purpose subroutines that are operated in real time by an executive program. This allows the DCU to provide total vehicle command and control including: tank pressurization and vent management, dynamic stability, sequencing, propellant consumption management, telemetry formatting, and pre-flight testing, as well as guidance and navigation functions.

PERFORMANCE IMPROVEMENTS

A number of performance improvements have been incorporated into Atlas/Centaur in order to meet increasing Intelsat V requirements without spacecraft apogee kick motor changes. These included an engine thrust increase (1,500 lb/eng), a zero-gravity parking orbit coast, earlier nose fairing jettison, and weight improvements. Additional changes are being incorporated to the Atlas/Centaur for the first Intelsat VA launch. These changes include adding a silver throat to the Centaur engine (+2.4 sec Isp), deleting the boost pumps, and using hydrazine for reaction control. The resulting capability for synchronous transfer is shown in Figure 6, and indicates a payload system weight capability of 4,900 lb for a near optimum transfer. Recently, the Intelsat Board approved an 80-inch Atlas stretch for the 1984 launches that will increase the Atlas/Centaur peformance capability to 5,300 lb.

Additional performance growth could be available as shown in Figure 7, by adding four Castor IV solid propellant motors and igniting two on the ground and two at altitude after the first two burnout. These are the same motors currently used with Delta and would result in a synchronous transfer capability for launching two full size Delta class 2,800 lb spacecraft on a single Atlas/Centaur using a Centaur tandem adapter (CTA) as indicated in Figure 8.

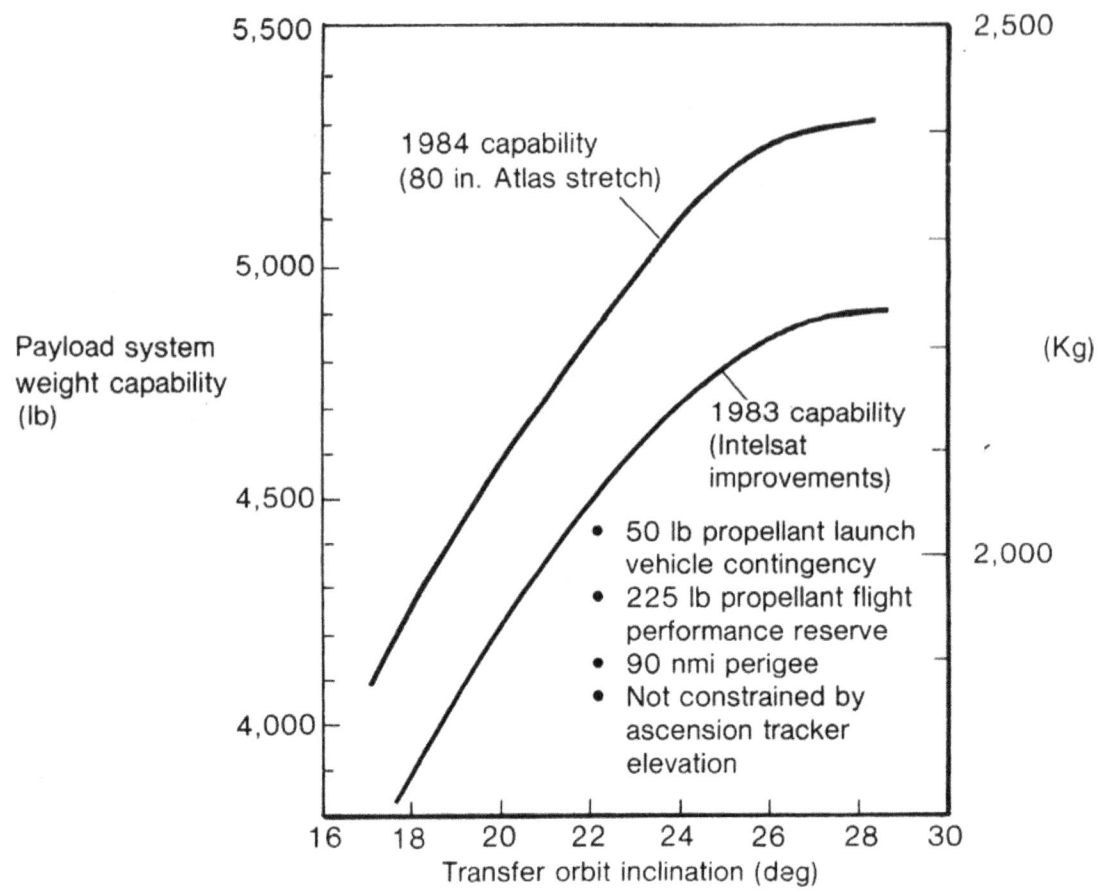

Figure 6. *Atlas/Centaur synchronous transfer performance.*

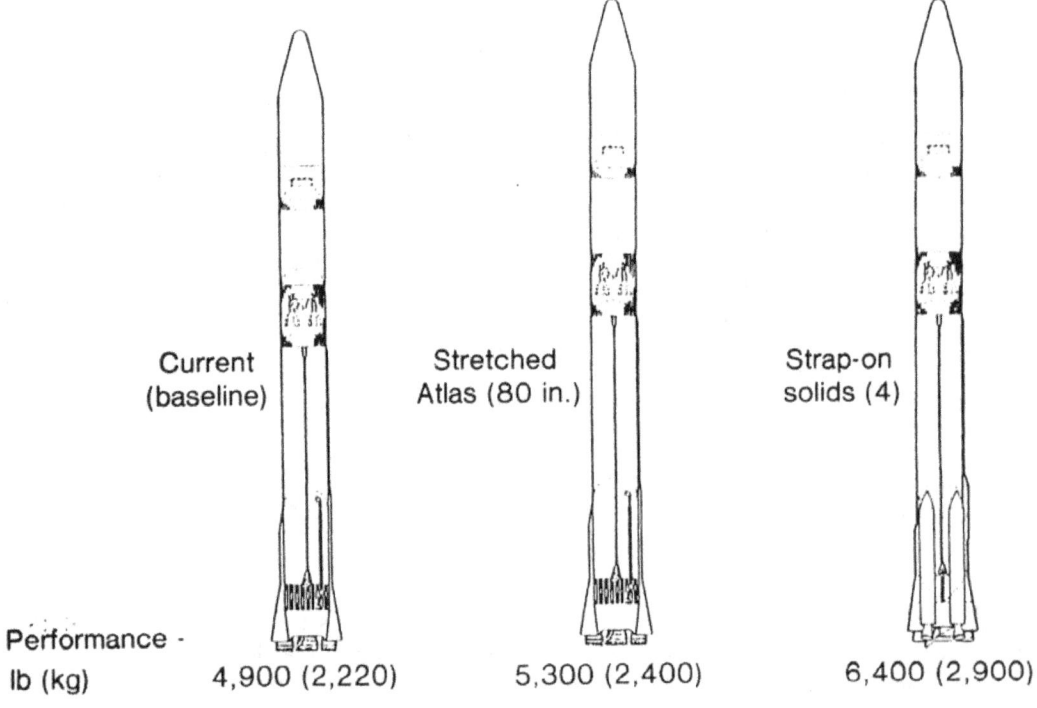

Figure 7. *Atlas/Centaur can provide additional performance growth.*

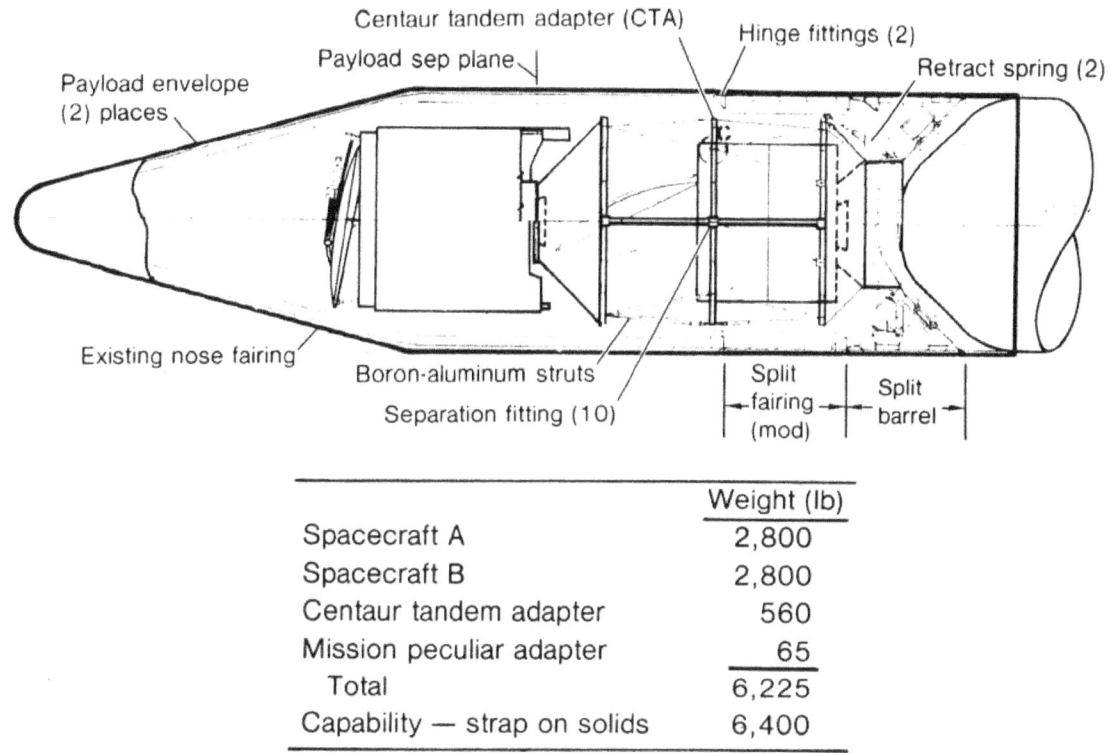

	Weight (lb)
Spacecraft A	2,800
Spacecraft B	2,800
Centaur tandem adapter	560
Mission peculiar adapter	65
Total	6,225
Capability — strap on solids	6,400

Figure 8. Dual spacecraft can be flown with Atlas/Centaur capability.

LAUNCH SCHEDULE

The current firm Atlas/Centaur launches are shown in Figure 9. Three Intelsat V launches, one COMSTAR and one Fleetsatcom launch are scheduled in 1981. COMSTAR 4 was successfully launched on 21 February 1981 with the second Intelsat V as the next scheduled launch in May.

Currently, Intelsat is considering accelerating the production schedule for the three Intelsat V-A vehicles in order to be able to launch them starting in mid-1983, and replacing these vehicles with additional Atlas/Centaurs for launch in 1984.

SHUTTLE/CENTAUR

General Dynamics Convair Division, under a number of separate contracts and company funded activities, has been studying the integration of the Centaur stage into the Space Shuttle since 1971. This activity culminated with a 1979 Centaur-in-Shuttle integration study which defined in detail the modifications to Centaur required for interface compatibility and mission safety.

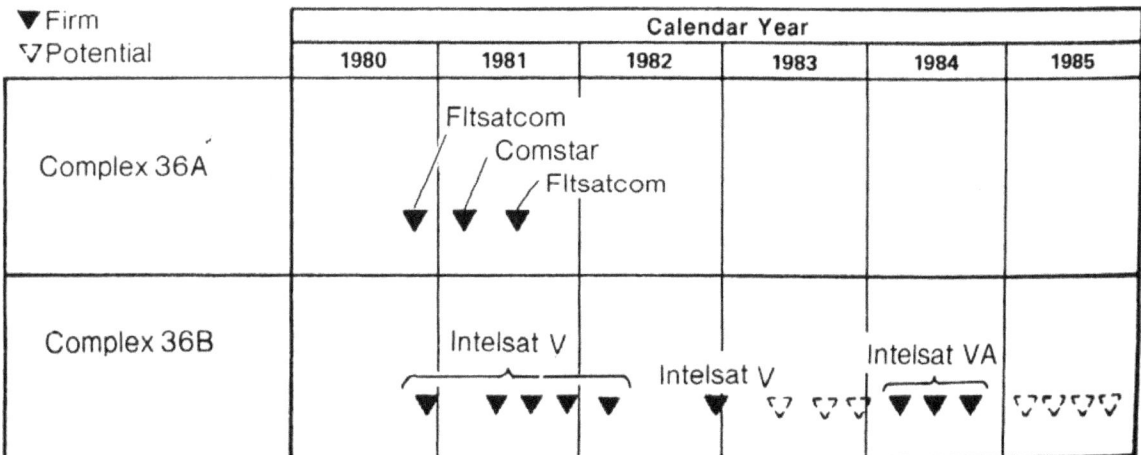

Figure 9. Atlas/Centaur launches are planned through the mid 1980s.

Shuttle modifications were defined by Rockwell International under the direction of NASA-Johnson Space Center (JSC). The ability to safely integrate Centaur-in-Shuttle was determined by JSC and Kennedy Space Flight (KSC) center personnel during Phase 1 safety reviews. All of these activities concluded that Centaur is ready for Space Shuttle integration (Figure 10).

During the last two months of 1980, NASA conducted a concentrated study of Shuttle upper stages to perform the Galileo and International Solar Polar (ISPM) missions. The conclusion of that analysis was the FY 1981 and 1982 budgeted resources would allow NASA to begin modification of Centaur for integration with the Shuttle so this powerful combination would be available for first launch in 1985. Shuttle/Centaur would satisfy the NASA planetary mission requirements and also be available for future commercial and national security missions. In January 1981, Dr. Frosch made the decision to recommend to Congress and the administration that NASA pursue this course of action.

CENTAUR MODIFICATIONS FOR SHUTTLE

The Centaur stage modifications required for Shuttle compatibility can be separated into two distinct areas: (1) interface compatibility and mission requirements and (2) safety considerations. For interface compatibility, the structural adapters must be modified; fill, drain, and vent systems are changed; a new LH_2 tank insulation blanket is added; and zero-g vent devices are required. Mission requirements dictate a TDRS compatible transponder and may require a star tracker for guidance update to meet mission accuracies if a number of Shuttle orbits are required prior to Centaur deployment.

Safety considerations required additional redundant valves and tank pressure transducers for propellant control, a new propellant dump system, reconfiguring of the helium purge system, and a new radio command link to inhibit engine firing after separation from the orbiter should a problem arise. These modifications still leave 95% of Centaur stage components unchanged.

WIDE BODY CENTAUR

The 1979 Centaur-in-Shuttle studies were based on using the current Centaur stage, which is 30 ft in length and 10 ft in diameter, and integrating it directly into the Shuttle. Performance requirements for the 1985 Galileo and ISPM missions dictated a significant increase in propellant weight that would be required to accomplish the missions. Since stretching the current Centaur in length only would result in inefficient use of the cargo bay, it was decided to stretch the LH_2 tank to 170 inches in diameter, thus utilizing the full orbiter bay available and increasing the length available for spacecraft as shown in Figure 11. The LO_2 tank diameter was held constant while a 30-inch cylindrical section was added. Centaur's propulsion system is unchanged; however, the forward equipment module and adapter diameter was increased to match the hydrogen tank diameter.

PERFORMANCE CAPABILITIES

Shuttle/Centaur performance capability significantly expands NASA's capability to perform solar exploration missions. Galileo can be launched in 1985 as a combined orbiter/probe using either a modified Type I trajectory (broken plane) or a Type II trajectory which arrives somewhat later. This is one of the most difficult

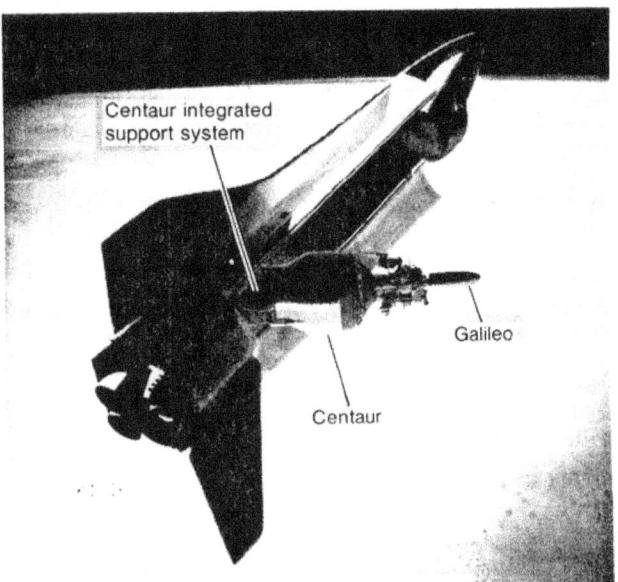

- Centaur integration & modifications defined by General Dynamics & NASA LeRC

- Shuttle modifications defined by Rockwell & NASA JSC

- Safe integration determined by NASA Johnson Space Center & Kennedy Space Flight Center

Figure 10. Centaur is ready for Shuttle integration.

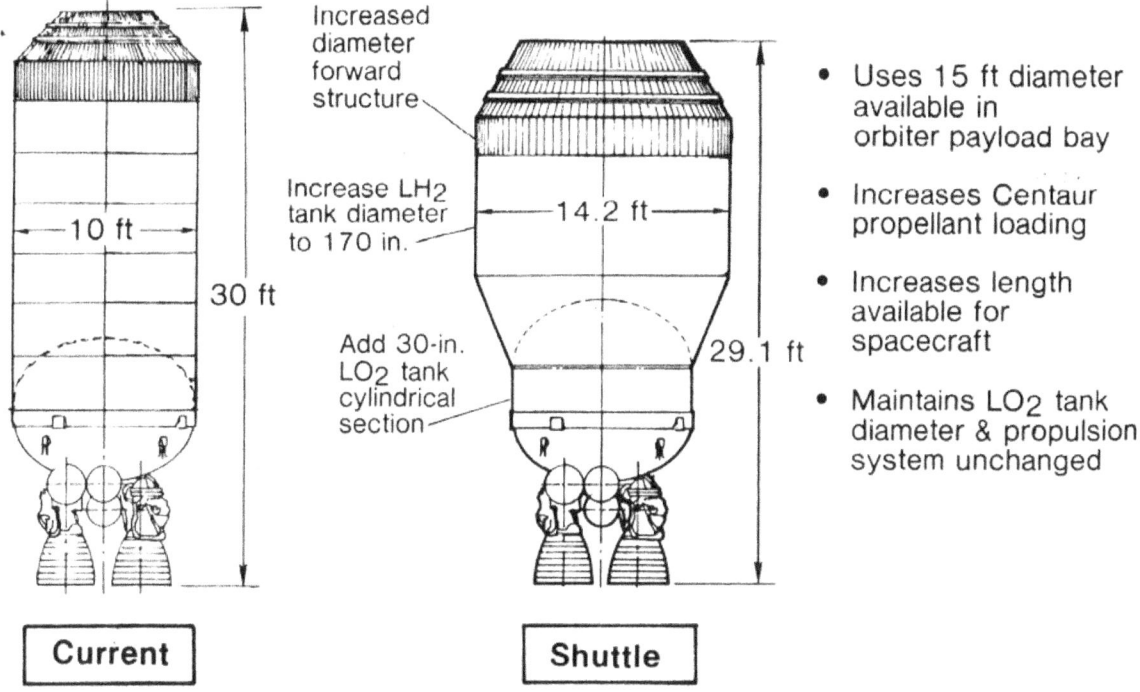

Increased diameter forward structure

Increase LH2 tank diameter to 170 in.

Add 30-in. LO2 tank cylindrical section

10 ft

30 ft

14.2 ft

29.1 ft

- Uses 15 ft diameter available in orbiter payload bay
- Increases Centaur propellant loading
- Increases length available for spacecraft
- Maintains LO2 tank diameter & propulsion system unchanged

Current

Shuttle

Figure 11. The wide body Centaur is a minimum modification to the current Centaur.

launch years for Galileo and could not be accomplished with any other upper stage, even by launching the orbiter and probe separately (Figure 12).

International Solar Polar can also be launched in 1985 with a single launch for both spacecraft and during the same launch opportunity as Galileo; however, a Star 48 upper stage is required. Future solar explora-

tion missions such as Venus Orbiting Imaging Radar (VOIR), Saturn Orbiter Probe, Uranus Orbiter Probe and Solar Probe will benefit from this improved capability.

Shuttle/Centaur will have a geosynchronous capability of 14,000 lb in orbit with the standard wide body configuration. This is nearly three times the pro-

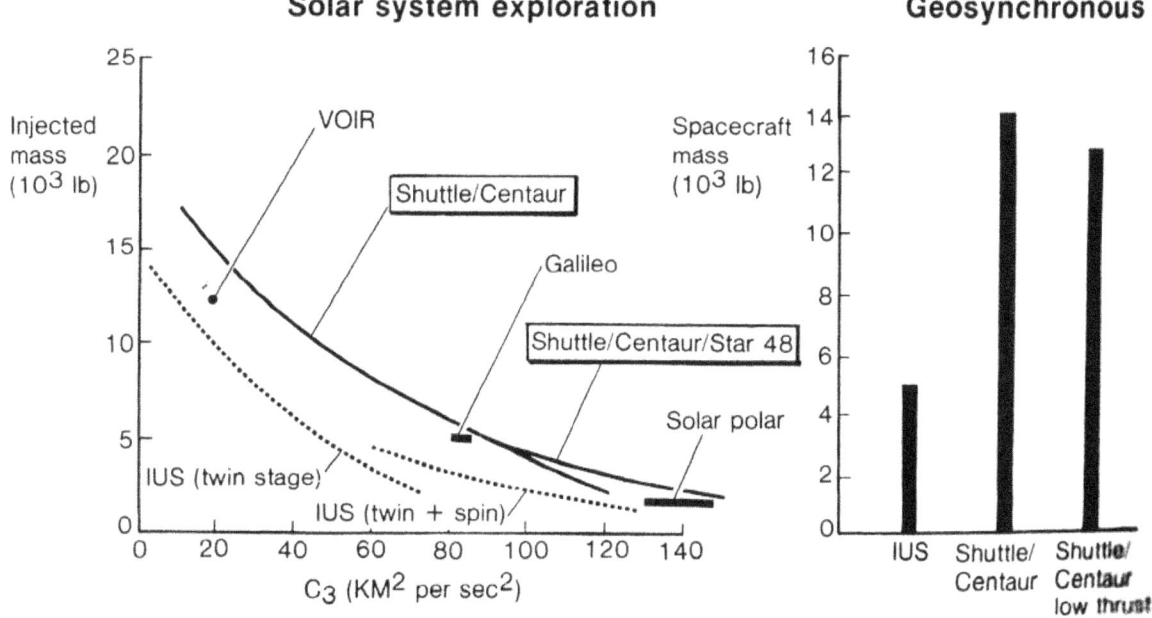

Solar system exploration

Injected mass (10^3 lb)

VOIR

Shuttle/Centaur

Galileo

Shuttle/Centaur/Star 48

Solar polar

IUS (twin stage)

IUS (twin + spin)

C_3 (KM^2 per sec^2)

Geosynchronous

Spacecraft mass (10^3 lb)

IUS Shuttle/Centaur Shuttle/Centaur low thrust

Figure 12. Centaur dramatically enhances Shuttle capability.

jected capability of Shuttle/IUS. For spacecraft with large deployed structures, Centaur engines can be operated in a low thrust mode (83,000 sec test time), which will allow the structures to be deployed and checked out prior to synchronous transfer. The geosynchronous capability for this type of mission exceeds 12,000 lb on orbit.

POTENTIAL MISSIONS

Some of the potential missions that Shuttle/Centaur could launch during the latter part of the 1980s include NASA's solar exploration missions Galileo, ISPM, VOIR, Solar Probe, and either an asteroid mission or comet mission. A geostationary platform demonstration flight, planned for 1987, could use Centaur capability in the low thrust mode. Large commercial communication satellites could also be launched either in single or tandem mode.

SUMMARY

Atlas/Centaur is a flight proven, reliable launch vehicle currently being uprated to deliver 5,300 lb to geosynchronous transfer orbit. Potential uprates include strap-on solid motors that could further increase this capability to 6,400 lb. This would allow launch of dual full-size Delta class spacecraft. Launches for Atlas/Centaur extend into the mid-1980s.

Shuttle/Centaur is ready for integration and now planned for a 1985 launch. Vehicle modifications required are minimal, and the stage has been determined safe by NASA. Geosynchronous orbit capability for this vehicle is 14,000 lb with a potential capability of 12,000 lb for a low thrust mission. Centaur will enhance NASA's solar system exploration capability with missions for this powerful combination extending through 1989 and beyond.

NASA Technical Memorandum 83795

A High Energy Stage for the National Space Transportation System

Andrew J. Stofan
Lewis Research Center
Cleveland, Ohio

Prepared for the
Thirty-fifth Congress of the International Astronautical Federation
Lausanne, Switzerland, October 7-13, 1984

A HIGH ENERGY STAGE FOR THE NATIONAL SPACE TRANSPORTATION SYSTEM

Andrew J. Stofan
Director
National Aeronautics and Space Administration
Lewis Research Center
Cleveland, Ohio 44135

Abstract

The Shuttle/Centaur is an expendable hydrogen/oxygen cryogenic upper stage for use with the National Space Transportation System. It is a modification of the existing Atlas/Centaur which has been used by NASA since 1966 to launch interplanetary and Earth orbital payloads for numerous organizations. Two configurations of the Shuttle/Centaur are being developed. Vehicle capability includes placing approximately 4500 kg (10,000 lb) in geostationary orbit, and initial applications will be for the interplanetary Galileo and Ulysses Missions in 1986. This paper discusses the Shuttle/Centaur development program, describes the configurations and performance, and indicates the unique integration and operations requirements related to the Shuttle. Design changes to the current Atlas/Centaur required for Shuttle operation are described here, and include those related to Orbiter cargo bay dimensions, environment, and safety considerations.

Introduction

Since the presentation of reference 1 at last year's 34th IAF Congress, there has been significant progress in the development, testing, and production of the Centaur upper stage for the National Space Transportation System (NSTS) (Figure 1). This paper will report on that progress, describe the configuration and performance of the two basic vehicles being developed, and discuss the unique integration and operations requirements related to the Shuttle.

The Centaur is an expendable high energy upper stage, using liquid hydrogen and oxygen propellants. It has been utilized for more than 20 years with Atlas and Titan boosters, and has launched 12 of the 15 successful planetary spacecraft for the United States, including all missions since 1962. Centaur has an overall operational reliability of 98% since 1971. Development of the Shuttle/Centaur capability was initiated in 1982 as a requirement for the Galileo and Ulysses (i.e., formerly International Solar Polar) Missions. The principal contractor is the General Dynamics Convair Division. Major contributions are provided by Honeywell, Teledyne, and Pratt & Whitney Aircraft. Project management is provided by the NASA Lewis Research Center.

First application of Shuttle/Centaur will be to launch the European Space Agency's Ulysses Spacecraft in May 1986 to Jupiter, where that planet's gravitational field will alter the spacecraft trajectory such that it will pass over the solar poles, exploring the out-of-ecliptic regions of the Sun for the first time. A few days later, the Jet Propulsion Laboratory's Galileo spacecraft will be launched on a less energetic trajectory to subsequently orbit and probe the atmosphere of Jupiter, and investigate its satellites.

Both of these initial missions will utilize the G-Prime version of the Shuttle/Centaur upper stage that fills about half of the Orbiter's 18.3 m (60 ft) cargo bay length (Figure 2). The second or G version of Shuttle/Centaur is about 3.0 m (10 ft) shorter, thus providing as much as 12.2 m (40 ft) for the payload length. Capability of the G vehicle is approximately 4500 kg (10,000 lb) to geostationary orbit. However, initial NASA planning for this upper stage also includes an interplanetary mission, the Venus Radar Mapper to be launched in 1988.

Performance

As developed for the Galileo mission, the Shuttle/Centaur G-Prime vehicle can inject a 2405-kg (5302-lb) payload into an Earth-escape hyperbolic trajectory with a single burn of the upper stage's main engines providing a C_3 energy of 80 km^2/sec^2 ($6.9 \times 10^{10}/ft^2/sec^2$). This is to be accomplished for launch from the Eastern Launch Site using a 241-km (130-nmi) circular, 28.5° inclined parking orbit provided by the Shuttle. The single Centaur burn is to occur no earlier than 45 min after separation from the Orbiter.

The principal option is a geostationary mission to be launched with a G vehicle from the same parking orbit. This mission profile results in a current spacecraft system weight capability of 4170 kg (9413 lb) to geostationary orbit with a zero degree inclination. Deployment from the Orbiter is assumed to be within 8 hr after liftoff, but separation can be delayed up to 84 hr after liftoff with corresponding performance degradation. The geostationary mission incorporates two burns of the Centaur main engines, as illustrated in Figure 3. The first burn occurs nominally 45 min after separation from the Orbiter. The second Centaur burn occurs after a 5-1/4-hr Hohmann transfer coast. Following spacecraft separation, the Centaur will execute a collision/contamination avoidance maneuver.

The 4270-kg (9413-lb) geostationary capability is based on the development status of the G vehicle at the March 1984 Preliminary Design Review (PDR). This weight includes a launch vehicle reserve of 85 kg (188 lb) for design maturity, which is equivalent to an additional 136 kg (300 lb) of spacecraft system weight. Since the PDR, launch vehicle system weights have decreased, improving payload capability.

In reference 1, mass and performance data were presented for the Shuttle/Centaur G-Prime, which has a propellant capacity of about 20,850 kg (46,000 lb). Performance data for the Centaur G (which has a propellant capacity of about 13,600 kg (30,000 lb)), were not presented. The Centaur G is appropriate for longer, heavier payloads requiring lower energies.

Table 1 provides current G-Prime vehicle mass summaries for the Galileo mission and for a more representative baseline mission with a characteristic velocity of 14.85 km/sec (48,710 ft/sec). For both of the missions shown, the current Shuttle lift capability precludes fully loading the Centaur tanks. The Shuttle/Centaur Galileo mission is a special case due to the commitment and preparation of a specific Orbiter to meet the Galileo requirements. Centaur masses and expendables for the Galileo mission are based on mission groundrules peculiar to that mission. The total loaded mass for the Galileo mission is 29,484 kg (65,000 lb) rather than the standard 27,442 kg (60,500 lb). The groundrules for the G-Prime baseline mission are as below.

Figure 4 shows Shuttle/Centaur G-Prime performance capability as a function of characteristic velocity. These data are also an update of the data of reference 1. The Centaur and CISS masses and groundrules are consistent with current Galileo parameters, but the Shuttle data are based on the current NASA standard Shuttle capability groundrules: (1) Shuttle cargo lift capability to a 241-km (130-nmi) circular orbit with a 28.5° inclination is 27,442 kg (60,500 lb) and (2) Shuttle chargeable weight is 1202 kg (2650 lb). The Shuttle chargeable weight consists of the Shuttle-supplied hardware which is necessary to fly the Centaur. The Centaur tanks must be offloaded when the total loaded mass, including spacecraft, exceeds 27,442 kg (60,500 lb).

Table 2 provides Centaur G vehicle mass summaries for the geostationary orbit mission and for a representative mission with a characteristic velocity of 11.85 km/sec (38,880 ft/sec). The Centaur tanks are fully loaded in these cases. Figure 5 shows the Shuttle/Centaur-G performance capability as a function of characteristic velocity. The Shuttle cargo groundrules are the same as for the Centaur G-Prime.

System Description

To meet cost, reliability, and schedule requirements, integration of the Centaur with the Shuttle is being accomplished with minimum modifications to the current upper stage vehicle and Orbiter. To a large extent, this is possible because of the Centaur Integrated Support System (CISS) (Figure 1). The CISS is to be installed in the Orbiter cargo bay and will provide compatible mechanical, electrical, and fluid interfaces between the Centaur and Orbiter. It will be returned to the launch site by the Orbiter for reuse. Installation and removal of the CISS from the Orbiter will have minimal impact on Shuttle schedules and weight. Weight permanently added to Orbiters Challenger and Atlantis to support Centaur is to be only 122 kg (268 lb).

Because the Shuttle is a manned vehicle, safety requirements are more stringent than for previous Centaur applications. Numerous redundant electronic and fluid system components are required on the Centaur and CISS. To minimize upper stage vehicle weight, these redundant computers, valves, and other components are to be mounted on the CISS where feasible, rather than on the Centaur. From lift-off to deployment from the Orbiter, the Centaur will be in a relatively quiescent state, with the CISS monitoring and controlling active Centaur

systems required for Shuttle safety. Although the CISS weight increases the total Shuttle lift requirement, its use reduces the Centaur stage dry weight which increases mission payload capability. Also, since the CISS is to be returned with the Orbiter for reuse, recurring costs will be reduced.

Centaur Structure for G-Prime Vehicle

The Centaur tank structure is pressure stabilized using the weight effective thin stainless steel tank technology developed initially by General Dynamics Convair Division. The Centaur tank consists of a liquid hydrogen and liquid oxygen tank joined by a common intermediate bulkhead. The liquid hydrogen tank is at the forward end of the vehicle and consists of a 4.32-m (170-in.) diameter cylindrical section closed by an ellipsoidal forward bulkhead and a 24° conical aft transition section that attaches to the LO_2 tank at its forward bulkhead/cylindrical section joint.

The liquid oxygen tank is formed by two ellipsoidal bulkheads of 3 m (120 in.) major diameter and 2.2 m (87 in.) minor diameter with a 78.75-m (31-in.) cylinder inserted between the bulkheads. The tanks will hold about 21,000 kg (46,000 lb) of propellants. The relative size of the tanks is determined by the desired engine burn mixture ratio of 5/1 (oxygen/hydrogen mass ratio).

The Centaur stage avionics are mounted on the forward adapter which consists of conical and cylindrical sections (Figure 2). The 33° conical section is a skin stringer aluminum alloy structure which is 119 cm (47 in.) long with a 4.32-m (170-in.) diameter at the base and a 2.7-m (108-in.) diameter at the forward end. The cylindrical section is 63.5 cm (25 in.) long and bolts to the liquid hydrogean tank forward ring. The cylindrical section skin is graphite-epoxy composite material.

At the rear of the Centaur, an aft adapter distributes CISS support loads into the Centaur oxygen tank. This is a 3-m (10-ft) cylindrical graphite/epoxy skin structure 28 cm (11.2 in.) long, which bolts to the liquid oxygen tank ring on the forward end and to the separation ring at the aft end.

The liquid hydrogen tank requires insulation that is effective in the atmosphere as well as in space. The liquid hydrogen tank forward bulkhead is insulated by a two-layer foam blanket under a three-layer radiation shield. The area between the forward bulkhead and the forward adapter is purged with helium during atmospheric operations to prevent condensation on the cold hydrogen tank walls. The same foam insulation is used on the liquid hydrogen tank sidewalls for prelaunch thermal control. Three radiation shields surround the foam insulation blanket on the sidewalls with the innermost shield acting as a sealed membrane to contain the helium purge. All other radiation shields are vented so they can be easily evacuated during ascent.

The liquid oxygen tank sidewall insulation is an extension of the liquid hydrogen radiation shield assembly, except all three shields are vented. The helium-purged foam blanket is not required for prelaunch insulation of the oxygen tank sidewalls and is omitted. The oxygen tank aft bulkhead is insulated by four vented radiation shields. All shield surfaces exposed to the Sun have a teflon surface with an under layer of vacuum deposited aluminum to achieve a low solar absorptance-to-emittance ratio for thermal control. A twin-skin vacuum intermediate bulkhead separates the two tanks. This is the same system that has been employed on all Centaur vehicles and yields very low heat transfer across the bulkhead.

The Centaur vehicle structural components will be designed to provide ultimate factors of safety greater than or equal to 1.40 while in the Orbiter bay and 1.25 after deployment.

CISS Structure for G-Prime Vehicle

The aft adapter on the Centaur stage mates to the deployment adapter on the CISS through the Lockheed Super*Zip separation ring, which is a dual pyrotechnic system. When the Super*Zip is fired, the ring is severed and a spring system thrusts the Centaur from the Orbiter at a velocity of 1/3 m/sec (1 ft/sec).

The deployment adapter transfers Centaur loads to the Centaur support structure during flight within the Orbiter and includes the rotation mechanism and the separation spring system. The basic structure of the 3-m (10-ft) diameter, 1.1-m (44-in.) high adapter is conventional aluminum skin-stringer construction. Just prior to deployment, the adapter rotates 45° around the two Centaur support structure trunnion pins. The deployment adapter supports the two fluid umbilical panels, the electrical umbilical panels, valve panels, deployment actuator fittings, and an avionics mounting shelf. It also provides a structure which helps support the two Centaur engines during Shuttle flight.

Centaur/Orbiter Structural Interfaces for G-Prime Vehicle

The Centaur vehicle and CISS attach structurally to the Orbiter at 8 points. Two aft retention pins on the CISS share the aft vertical load, and two forward retention pins on the CISS share the vertical load with the aft pins, and react all axial loads from the Centaur with its payload. CISS keel pins on the forward adapter share the lateral loads. Two sill pins on the forward adapter share the vertical loads with the CISS pins.

G-Vehicle Differences for Centaur and CISS Structures

The Centaur tanks and CISS for the G vehicle are basically the same construction as for G-Prime. The G tanks are smaller holding 13,500 kg (29,600 lb) of propellant for a 6/1 mixture ratio. The overall vehicle is 6.1 m (20 ft) long vs. 9.0 m (29.6 ft) for G-Prime (Figure 2). The CISS is shorter to be compatible with the shorter G vehicle.

Propulsion Systems

The Shuttle/Centaur main propulsion system consists of two Pratt & Whitney RL-10 regeneratively cooled hydrogen/oxygen engines using an expander cycle. The Shuttle/Centaur G-Prime utilizes RL-10-3-3A engines operating at a 5/1 nominal mixture ratio and 73,400 N (16,500 lb) nominal thrust each with a specific impulse of 446.4 sec. The

Shuttle/Centaur G uses RL-10-3-3B engines at 6/1 mixture ratio, 66,700 N (15,000 lb) thrust each with 440.4 sec specific impulse. The engines operate at constant thrust and are capable of multiple starts after long coast periods in space, as demonstrated in previous missions. Each engine contains a hydraulic gimbal actuation system which is powered by the turbopump assembly.

Prior to engine ignition, the propellants must be settled to the bottom of the tanks to provide the liquid propellants at the sumps for engine operation. The settling thrust is provided by the auxiliary propulsion system. After the propellants are settled, the propellant turbopumps are chilled and primed before each engine operation to prevent pump cavitation during the engine start transient. Hydrogen and oxygen are flowed concurrently through the appropriate pumps during the prestart cycle prior to each engine ignition to chill them to their required temperatures.

The auxiliary propulsion system for attitude control during coast phases and for settling the propellants prior to the main burn consists of twelve 27 N (6 lb) hydrazine monopropellant thrusters. Four thrusters are oriented axially and provide settling thrust. The other eight thrusters provide pitch, yaw, and roll control. The thrusters are fed from a positive expulsion bottle capable of holding up to 77 kg (170 lb) of hydrazine. An additional bottle may be added as required. The hydrazine thrusters are manufactured by Hamilton Standard.

Fluid Systems

The Shuttle/Centaur fluid systems include the main engine propellant supply, the cryogenic vent fill/dump, hydrazine, reaction control, hydraulic, and pneumatic systems. The fluids include liquid hydrogen, liquid oxygen, gaseous hydrogen, helium, hydrazine, and hydraulic fluid.

The main engine propellant supply system provides propellants to the engine (as described earlier) with the net positive suction head (NPSH) required by the engine turbopumps. The required NPSH is provided prior to engine start by pressurizing the vehicle propellant tanks with helium. During a burn the oxygen tank is also pressurized with helium. In the case of the hydrogen tank, gaseous hydrogen from the main engines is bled back into the hydrogen tank to maintain pressure and conserve helium. Flexible insulated feed ducts deliver propellants from the tanks to the engine turbopumps. To prevent inadvertent opening of the propellant inlet shutoff valves, a parallel set of pyro valves and solenoid valves upstream of the pneumatically actuated control solenoid valves provides the two-failure-tolerance dictated by Shuttle safety requirements. The pyro valves will be fired open shortly before the first main engine burn.

The main propellant tanks are filled through the CISS after the Centaur and spacecraft are installed in the Orbiter prior to launch. Unlike a solid fueled stage, the Centaur propellant tanks can be dumped in case of a Shuttle abort after launch. For compatibility with all Shuttle abort modes, the fill/dump system has been sized to provide single-failure-tolerant propellant dump capability within 250 sec. Propellant dump is performed by using pressurized helium stored in bottles on the CISS and the Centaur to force the hydrogen and oxygen from the tanks overboard through the CISS. After the propellants are dumped, the tanks are purged with helium. Capability for abort dumping with the Centaur will allow the Orbiter to land with about 10,000 kg (22,000 lb) in the cargo bay rather than the 29,500 kg (65,000 lb) that otherwise would be in the bay with a fully loaded Centaur.

To prevent large outages (residuals) of either hydrogen or oxygen at the end of flight, Centaur has an active propellant utilization system. The system takes propellant level data obtained from capacitance probes in the propellant tanks and adjusts engine mixture ratio during a burn to maintain the nominal propellant mixture ratio in the tank. Since the main impulse propellants are cryogenic, heat leaks into the tanks cause the propellants to boil and the tank pressures to rise. A vent system is required to control pressure in the tanks. For each propellant tank, pressure is controlled by a parallel set of valves, one mechanical self-regulating vent valve with solenoid lockup capability and one solenoid operated valve. During the time the Centaur is in the Orbiter bay, propellants are vented overboard through the CISS at various locations on the Orbiter sidewalls and aft section.

The liquid hydrogen tank has a thermodynamic vent system for control in a zero-g environment. The thermodynamic vent system has an electrically driven pump to circulate hydrogen over heat exchanger coils and mix the bulk hydrogen. The oxygen tank is not expected to require venting in orbit providing the bulk fluid is well mixed. A pneumatically operated jet pulse mixer assures that the liquid oxygen is well mixed.

For the G-Prime Centaur, two helium bottles on the Centaur and 18 on the CISS contain the helium necessary to meet the pressurization, dump, purge, and pneumatic valve operation requirements. The helium bottles are Kevlar-overwrapped metal-lined spheres. The Centaur bottles are 66 cm (26 in.) in diameter and the CISS bottles are 56 cm (22 in.) in diameter. The G version of the vehicle has an additional 56 cm helium sphere on the Centaur and 12 spheres of 56 cm diameter on the CISS.

The auxiliary propulsion system consists of the twelve 27 N (6 lb) thrusters mentioned earlier. The tank is a positive expulsion type with a 77-kg (170-lb) capacity. The feedlines from the tank to the motors are heated to prevent freezing due to their proximity to cryogenic propellants and exposure to space conditions. The feedline joints are welded to ensure an absolutely leak-proof contamination-free system. For safety reasons, there are pyro valves in the hydrazine tank inlet and outlet lines to provide positive isolation of the hydrazine tank. A downstream set of parallel pyro valves and solenoid valves provides two-failure-tolerance against inadvertent thruster operation. Like the pyro valves in the main propulsion system, the pyro valves in the auxiliary control system will not be fired until the Centaur has reached a safe distance from the Orbiter, about one hundred meters (300 ft).

Thrust vector control is provided during main burns by gimbaling the RL-10 engines using hydraulic actuators. Each engine is driven by two

closed-loop, servo controlled actuators. The systems on each engine are independent with an engine-driven main hydraulic pump and an electric-motor-driven recirculation pump. The hydraulic system is inactive in the Orbiter bay except for operation of the recirculation pump during lift-off and abort landing, and intermittent operation on orbit for thermal conditioning of the hydraulic system components.

Avionics and Electrical Systems

The avionics system on the Shuttle/Centaur/CISS performs or controls the guidance and navigation, control, sequencing, propellant utilization, vent and pressurization, instrumentation, and telemetry functions. The system on the Shuttle/Centaur is very similar to that currently used on the Atlas/Centaur, and includes the Teledyne Systems Company digital computer unit (DCU), a Honeywell Inc. inertial measurement group (IMG), a sequence control unit (SCU), a servo inverter unit (SIU), a dual-failure-tolerant arm/safe sequencer (DUFTAS), and other avionics equipment.

The DCU is a 16,384-word, 24-bit, random-access, core memory computer. The IMG provides the DCU with a measurement of vehicle accelerations, using a four-gimbal, gyro-stabilized platform which supports three orthogonal, pulse-rebalanced accelerometers. The guidance and navigation function for the Centaur is performed by the DCU, which takes the vehicle accelerations, integrates them appropriately, computes position and velocity, and generates required steering signals from the guidance algorithm.

Centaur main engine thrust vector control and coast phase attitude control are performed based on vehicle attitude errors received by the DCU from the IMG. The DCU computes desired engine actuator commands which are then sent from the DCU to the servo inverter unit. The SIU differences the command from a position signal from the engine actuator feedback transducer. This difference is power amplified and applied to the engine actuator servo valve. During a coast phase, attitude control signals are generated by DCU computations, again using input from the IMG. The sequence control unit receives input from the DCU which activates relays in the SCU which cause the appropriate attitude control thrusters to fire. Until the deployment is complete and Centaur is a safe distance from the Orbiter, a timing mechanism (DUFTAS) provides dual-failure-tolerance against inadvertent initiation of all Centaur hazardous functions, including both main engine and attitude control engine firings.

In addition to controlling the attitude control thruster firings, the DCU provides all sequencing commands to vehicle systems and required discrete commands to a spacecraft. These commands are routed to the SCU which activates relays implementing the discretes. Logic in the DCU minimizes main impulse propellant residuals by actively controlling engine mixture ratio. As mentioned earlier, capacitance probes in the propellant tanks sense propellant levels. These data are used by the DCU to determine the proper engine mixture ratio to minimize residuals. The DCU also monitors and controls the pressurization of the Centaur main propellant tanks according to a predetermined schedule, minimizes helium usage, and provides failure detection and corrective action for the redundant tank pressurization components. In addition to the above functions, the DCU with ancillary equipment manages the instrumentation/telemetry function to provide a data stream through the Orbiter while the Centaur is in the bay or in Orbiter proximity. After deployment and separation, the Centaur switches to the Tracking and Data Relay Satellite (TDRS) link which transmits data to the White Sands Ground Tracking Station.

CISS avionics provide the dual-failure-tolerant capability to meet Orbiter safety requirements. Five identical control units (computers), with majority voting, provide the redundancy to meet these requirements. All electrical interfaces between the Centaur and the Orbiter are provided by the CISS. The CISS also provides computer control prior to separation for operational sequencing of all systems requiring multi-failure-tolerance, propellant tank vent and pressurization control, electrical and power control, instrumentation and telemetry pyrotechnic control for Centaur separation from the CISS, and Centaur propellant-level tanking indications. The system is designed to minimize the Centaur avionic interfaces with the Orbiter.

Electrical power for the Centaur is provided by a silver-zinc battery of 150 amp-hr, which is switched on just prior to Centaur deployment. Additional batteries can be provided for longer missions. Backup CISS electrical power is provided by two 375 amp-hr silver-zinc batteries, redundant with Orbiter power to provide two-failure-tolerant power to the CISS/Centaur in the Orbiter-attached mode.

Software System

The software system includes Centaur vehicle software, CISS software, and extensive ground computer software. The Centaur and CISS software is modularized such that each module performs a unique and manageable segment of instructions which are individually coded, checked, and documented. This modular concept has been successfully used in the Atlas/Centaur program and provides the necessary flexibility and reliability for the rapid assembly of the software required to fly a wide variety of missions. The vehicle DCU software includes not only the software required to perform the functions described in the avionics section, but also software which supports ground checkout and launch operations. The flight software uses backups that are as forgiving as possible of hardware failures.

The CISS software supports ground testing, ground and launch support operations including tanking, predeployment testing and operations of the Centaur and the CISS, and prelanding and post-landing operations, including Shuttle abort. As stated earlier, the major purpose of the CISS software is to control in a one-or-two-failure-tolerant system all safety related functions while the Centaur vehicle is attached to the Orbiter. The CISS software is designed to relieve the burden which would otherwise be placed on the Centaur vehicle avionics or the Shuttle crew to control all safety related functions in a two-failure-tolerant mode. The ground computer software includes extensive capability for ground testing of the Centaur and CISS systems, the tanking operation, and prelaunch and launch operations.

Test Program

Ongoing development, qualification, and validation tests, together with previous experience with cryogenic Atlas and Centaur vehicles, will result in a low-risk program. Development testing for Shuttle/Centaur is planned to provide early solutions to design problems and to identify key characteristics of hardware and software. Component and/or subsystems will be tested in progressive stages to ensure earliest recognition of possible problem areas.

A limited amount of structural development testing will be necessary for Shuttle/Centaur. These development tests will consist initially of material elements and welding tests, separation tests, tank insulation tests, Centaur support structure load tests, and tests due to the new launch and landing environment. These new environmental tests include acoustics, vibration, Centaur loads, and modal surveys. Fluids and mechanisms systems for Shuttle/Centaur are based on Atlas/Centaur (i.e., Centaur D-1A) technology except where Shuttle imposes additional requirements. The development test will cover mechanisms to deploy the Centaur from the Orbiter, propellant dump capabilities, and fluid interfaces with the CISS.

Avionics and software on the Shuttle/Centaur also evolved from the Atlas/Centaur systems with all changes required because of the additional Shuttle requirements, such as safety, operations sequence, different-shaped vehicle, and platform torquing functions. Although most ground avionics will be the same as used on Atlas/Centaur, a few additional items of test equipment or test sets will be required to test the Centaur avionics and software. A functional integration test will verify that the vehicle with the CISS will meet vehicle-level performance requirements. The initial phase of functional integration consists of tests on each functional subsystem using the other systems as necessary to support the tests. These individual subsystem tests will concentrate on detailed performance and margins of each individual subsystem.

Factory testing of the Shuttle/Centaur first article will be similar to current practice. New tasks include checkout of redundant systems, vehicle-to-CISS interfaces, and combined Centaur/CISS system tests. Since the CISS is a new item, the test philosophy will be similar to Centaur testing using a CISS simulator. Early qualification testing will reduce cost and increase reliability. Component qualification testing for Shuttle/Centaur is expected to identify any design problems before system and subsystem-level testing. All components will have successfully completed functional checkout and acceptance testing including burn-in (if required) before qualification testing. Environmental qualification test requirements will comply with specifications. All newly designed components will be qualified to ensure full compliance with Shuttle requirements. Validation tests on the Centaur and CISS will increase reliability. These tests will be performed at the launch site.

Upon completion of the developmental phase of the test program, the structural test Centaur/CISS will be refurbished and used as a Pathfinder test vehicle. This Pathfinder test vehicle will be delivered to Florida for use at the Eastern Launch Site (Cape Canaveral Air Force Station and NASA Kennedy Space Center) to accomplish physical fit checks, along with handling and facility processing procedure checkouts. These Pathfinder test vehicle operations will be performed prior to the flight Centaur vehicle operations in each of the Eastern Launch Site facilities. The physical fit check with an actual Shuttle orbiter vehicle will be accomplished with the actual flight Centaur scheduled to perform the Galileo mission.

To conclude verification of designs, some first article flight hardware will also be utilized. An acoustics test will be conducted on the flight forward adapter with simulated avionics packages. A rotation/separation test will be conducted using the flight Centaur Support Structure (CSS), flight deployment adapter, separation ring, and the aft flight adapter.

Systems Integration Laboratory (SIL)

To support the integration of the Centaur vehicle as a high energy upper stage vehicle for the Shuttle program, a system to simulate and emulate the Shuttle/Centaur avionic flight system and its supporting ground control and checkout equipment has been developed. This system has been designated the Systems Integration Laboratory (SIL)[2]. The SIL is composed of integrated simulators that form a composite control system complement to the Centaur airborne and avionic support equipment. It provides an off-line capability to verify the system design of the Centaur airborne support equipment and the Centaur avionic flight systems. In addition, it provides a realistic medium for the development and integration of ground checkout and airborne control software programs.

Each simulator is composed of prototype hardware, where feasible, to maximize configuration likeness. Where emulated flight or ground hardware is used, it provides physical characteristics (loads, signals, etc.) equivalent to those of the flight hardware. The SIL is so designed that after it is used for integrated laboratory testing of the Centaur it will be transported to Kennedy Space Center, Florida, to verify the operational readiness of the Shuttle/Centaur prelaunch and launch complexes. The individual simulators will be mounted on a structure for handling purposes. This combination simulator will be utilized at the Eastern Launch Site for electrical and electronic functional checkouts of the facilities. These checkouts will be completed prior to use of the facilities by the first flight Shuttle/Centaur vehicles.

Manufacturing Status

The status of the G-Prime hardware, as of July 1984, will be described here. As of this date, fabrication of hardware for the G vehicle has not started. In both instances the first vehicle fabricated and assembled will be dedicated as a test vehicle with the remaining vehicles destined to support the space missions.

Fabrication and assembly of the test CSS started in February 1984 and was completed in July. The test tank fabrication/major weld started in January 1984 and was completed in May. This in-

cluded installation of internal mechanical hardware. Following this, the test tank underwent a high pressure leak check in June 1984. The test tank was then moved to the final assembly buildup area. The fabrication and assembly of the test forward and aft adapters started in February of 1984 and were also completed in July. The forward and aft test adapters will be installed on the test tank at the final buildup area in August 1984. The test deployment adapter assembly started in May of 1984 and was completed in mid-July. This adapter was then shipped to NASA's Goddard Space Flight Center via a Super Guppy aircraft to support a 3-wk acoustic test in late July/early August.

As stated above, this first article hardware was fabricated to be used in support of structural verification testing followed by a Pathfinder program. The Structural Test Program will start in August 1984 and is scheduled to be completed in March 1985. The tests will consist of the following: A CSS stiffness test, insulation/helium purge test, static load test, cargo modal survey test, and ending with a tank modal survey test. Following completion of the Structural Test Program the test hardware will be shipped to NASA's Kennedy Space Center (KSC) to verify interfaces between the test vehicle and the KSC facilities such as Complex 36A, Vertical Processing Facility (VPF), and Launch Complex 39. In addition, personnel not familiar with handling thin-walled pressure stabilized vehicles will be provided the opportunity to familiarize themselves prior to actual handling of the flight vehicles.

National Space Transportation System Operations

Ground Processing

The Shuttle/Centaur vehicles will be checked out during ground processing (Figure 6) at the Eastern Launch Site (ELS) in Florida prior to the launch of each vehicle in the Shuttle. Each Centaur and CISS will arrive at ELS on the NASA Super Guppy aircraft. The Centaur and CISS are to be transported to Hangar J for general inspection. The CISS will be transported to Complex 36A where it is to be installed and subjected to subsystem functional checkouts. The Centaur vehicle is then to be transported from Hangar J and installed on the CISS. Additional checkout and verification tests will be performed on the Centaur/CISS assembly prior to an actual cryogenic tanking which will be accomplished during a Terminal Countdown Demonstration Test. Upon completion of that test, the vehicle tanks are to be drained and purged. Finally, reaction control system propellants will be loaded into the Centaur hydrazine tanks.

The Centaur/CISS Assembly is to be transported from Complex 36A to either the Vertical Processing Facility (VPF) or the Shuttle Payload Integration Facility (SPIF). The operations at either of these integration facilities will involve mating a spacecraft to the Centaur and verifying interfaces between the Centaur/Spacecraft and the Centaur/Orbiter. The Centaur-to-Orbiter electrical interface is to be verified using the Cargo Integration Test Equipment at either of the two integration facilities (VPF or SPIF). The final integrated test prior to transport to the launch pad will be the Centaur Cargo Element (Centaur, CISS, and Spacecraft) End-to-End Test. This test encompasses all of the control and monitoring centers and is a test of the telemetry and command links.

The Centaur Cargo Element will be installed in the Multi-use Mission Support Equipment (MMSE) canister and transported to Launch Complex 39 where it is to be transferred to the Rotating Service Structure. After the Orbiter is mounted on the launch pad, the Rotating Service Structure will be rotated to the Orbiter for cargo installation. The Centaur Cargo Element is then to be installed into the cargo bay and all connections, leak checks, and functional tests finalized. A test will be performed to verify the interfaces between the Orbiter and the Centaur Cargo Element. An end-to-end test will also be performed to verify the communication links between the Orbiter and the control and command centers; and a dry and wet countdown demonstration test completed to verify cryogenic loading and mission readiness. The final countdown, concluding in the launch of the Shuttle, includes cryogenic tanking.

The Shuttle Orbiter, of course, is to return to ELS after successful deployment of the Centaur and Spacecraft. The Orbiter will be rolled into the Orbiter Processing Facility (OPF) where the payload bay doors are to be opened and the CISS removed. The CISS is then to be returned to Hangar J for refurbishment and support of a subsequent Shuttle/Centaur mission.

Flight Operations

During ascent and Orbiter-attached operations, primary flight operations control will rest with Johnson Space Center's Houston Mission Control Center (MCC-H). The Centaur Payload Operations Control Center (CPOCC) located at the ELS will monitor Centaur systems and provide appropriate GO/NO GO decisions for Centaur deployment phases to MCC-H. After separation, when the Orbiter has maneuvered out of the zone of safety around Centaur, CPOCC will continue to monitor Centaur. CPOCC monitoring continues through Centaur burns, spacecraft separation, and Centaur post-separation maneuvers. Actual CPOCC "control" of Centaur prior to separation from the Orbiter is limited to initiation of pre-programmed system checks, initiation of navigation updates, GO/NO GO decisions for rotation and separation from the Orbiter, data monitoring and recording, and evaluation of anomalies. Since there is no uplink to Centaur after separation from the Orbiter, during that period CPOCC is limited to data monitoring and recording, and evaluation of anomalies.

The sequence for rotation and separation of Centaur from the Orbiter requires several discrete actions by the Orbiter crew. After reorienting the Orbiter, inhibiting its primary maneuvering system, and receiving a "GO" for rotation, the mission specialist engages the rotation system clutches, releases the payload latches, and initiates the "ROTATE" sequence. Following successful rotation, initiation of the "COMMIT" sequence switches Centaur to internal power, activates a telemetry transmitter check, and configures the fluids systems for separation. Upon receipt of the "GO" for

separation from the ground, Orbiter maneuvering is again inhibited, and the mission specialist arms and fires the separation system.

Centaur's maneuvering system is inhibited until 5 min after Orbiter separation. The main engine system is inhibited until 45 min have elapsed after separation. Depending upon mission requirements, Centaur performs one or two main engine "burns" to achieve spacecraft trajectory and positioning. Following spacecraft separation, Centaur performs a deflection maneuver, vents residual propellants, and terminates operations.

Abort Operations

Orbiter malfunctions early in the ascent may dictate Return to Launch Site (RTLS) or Trans-Atlantic Abort Landing (TAL) abort modes. In these events, Centaur propellants are to be dumped according to an Orbiter-programmed sequence activated by a switch on the Orbiter's abort panel. Centaur propellant dumps for TAL Avoidance (TALA) and Abort Once Around (AOA) modes are initiated by mission specialist program execution from the Orbiter keyboard. An Abort To Orbit (ATO) situation may not require a Centaur propellant dump and may allow deployment of Centaur from a lower Earth orbit. If a Centaur dump is necessary, the mission specialist controls it from the keyboard as in the TALA and AOA modes.

An Abort From Orbit (AFO) that occurs with Centaur still on board the Orbiter will require an Orbiter maneuver to settle Centaur's propellants before the mission specialist initiates the dump cycle from the keyboard. For all abort modes except RTLS, residual Centaur propellants (remaining after the dump cycle) can be dumped during an Orbiter propellant settling maneuver or during the Orbiter's re-entry burn.

Conclusions

The new Shuttle/Centaur cryogenic stage will provide a capability, not currently available with solid rockets, to place large spacecraft into Earth orbit and high energy trajectories to the planets. With as much as 12.2 M (40 ft) of Orbiter cargo bay length available for spacecraft and a projected capability of 4500 kg (10,000 lb) to geostationary orbit, the addition of the Centaur to the fleet of Shuttle upper stages offers attractive mission possibilities. Shuttle/Centaur development is based on modification of the reliable Centaur D-1A currently used with Atlas boosters, and is progressing satisfactorily towards initial 1986 launch availability.

References

1. Spurlock, Omer F.; "Shuttle/Centaur - More Capability for the 1980's," IAF Paper 83-18, October 1983.

2. Gordan, Andrew L.; "Role of Simulation and Emulation in Development of Shuttle-Centaur," NASA TM-83326, July 1983.

TABLE 1

SHUTTLE/CENTAUR G-PRIME VEHICLE MASS SUMMARY (KG)

	BASELINE*	GALILEO
TOTAL LOADED MASS	27,442	29,484
TOTAL SUPPORT MASS	4,162	4,343
TOTAL VEHICLE MASS	23,281	25,141
SPACECRAFT SYSTEM MASS	1,357	2,561
CENTAUR TANKED MASS	21,924	22,580
CENTAUR JETTISON MASS	3,125	3,119
CENTAUR DRY MASS	2,605	2,605
CENTAUR RESIDUALS	277	277
FLIGHT PERFORMANCE RESERVE	81	99
LAUNCH VEHICLE CONTINGENCY	162	138
CENTAUR EXPENDABLES	18,799	19,461
PROPELLANTS	18,776	19,438
MAIN IMPULSE	18,549	19,211
OTHER	227	227
HYDRAZINE	22	22
HELIUM	1	1

*14.85 KM/SEC (48,710 FT/SEC) CHARACTERISTIC
VELOCITY.

TABLE 2

SHUTTLE/CENTAUR G VEHICLE MASS SUMMARY

	BASELINE*	GEO
TOTAL LOADED MASS	25,919	25,422
TOTAL SUPPORT MASS	4,139	4,139
TOTAL VEHICLE MASS	21,780	21,283
SPACECRAFT SYSTEM MASS	4,811	4,266
CENTAUR TANKED MASS	16,969	17,017
CENTAUR JETTISON MASS	3,537	3,642
CENTAUR DRY MASS	3,098	3,097
CENTAUR RESIDUALS	235	357
FLIGHT PERFORMANCE RESERVE	120	103
LAUNCH VEHICLE CONTINGENCY	84	85
CENTAUR EXPENDABLES	13,432	13,375
PROPELLANTS	13,410	13,319
MAIN IMPULSE	13,229	13,151
OTHER	181	168
HYDRAZINE	21	54
HELIUM	1	2

*11.85 KM/SEC (38,880 FT/SEC) CHARACTERISTIC
VELOCITY.

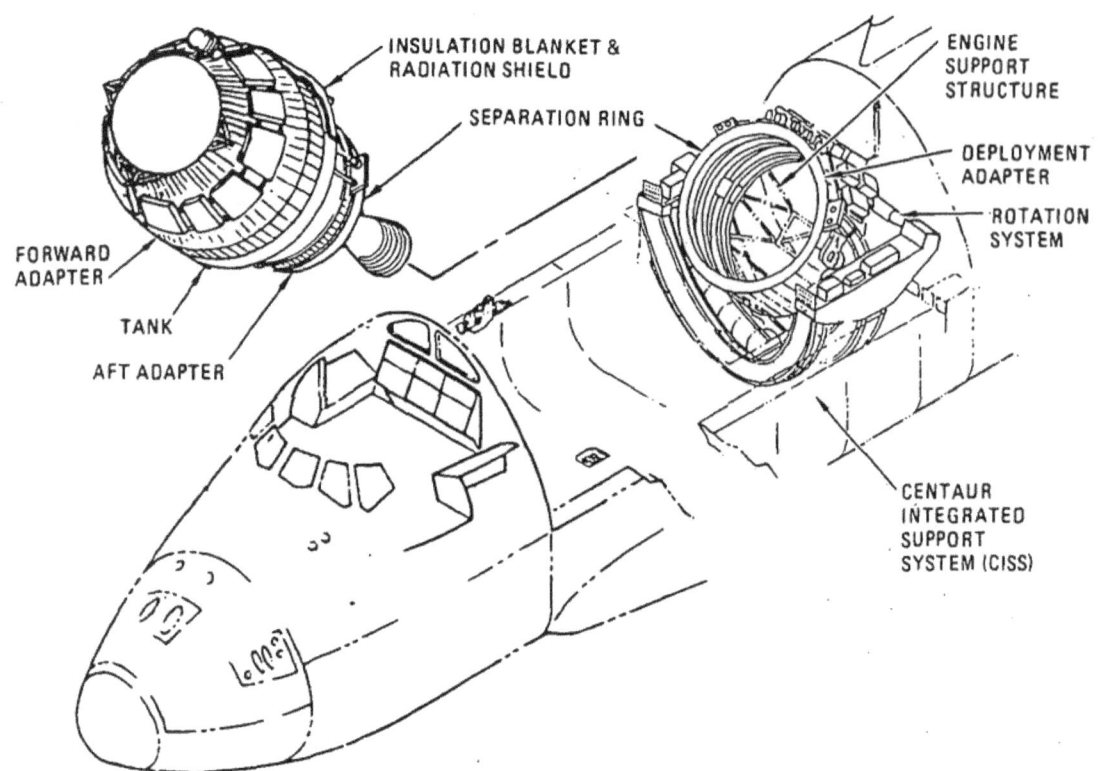

Figure 1. - Shuttle/Centaur system.

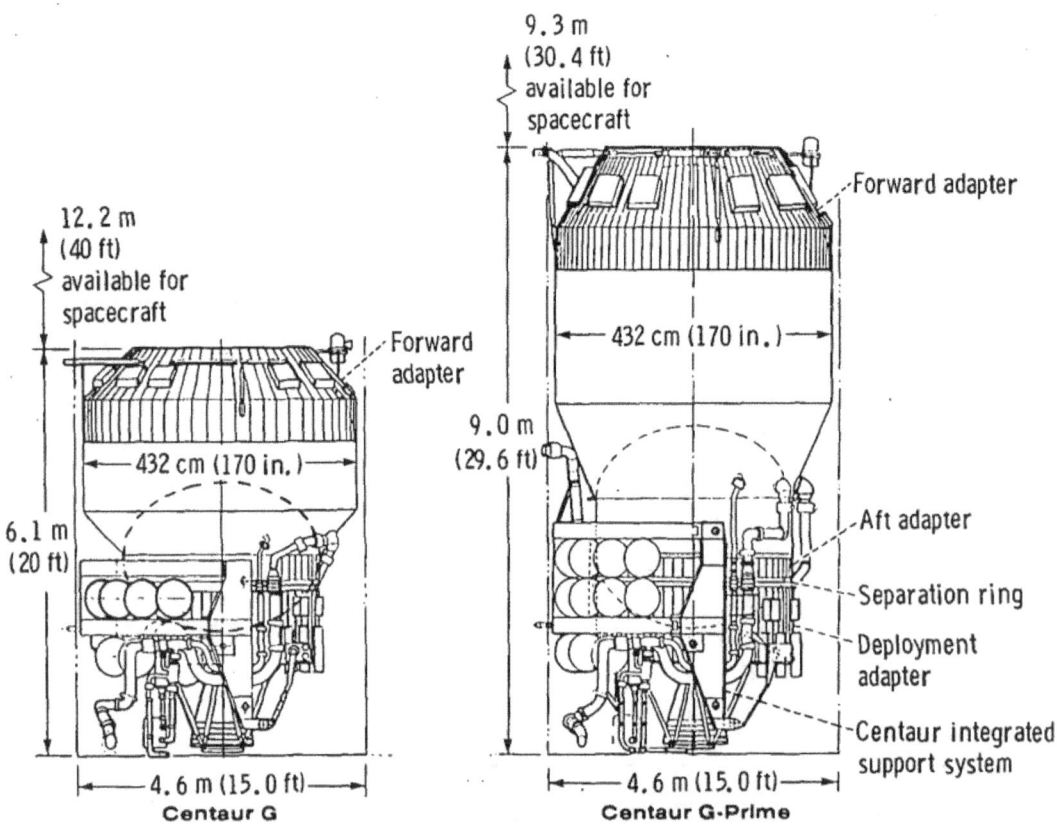

Figure 2. - Shuttle/Centaur configurations.

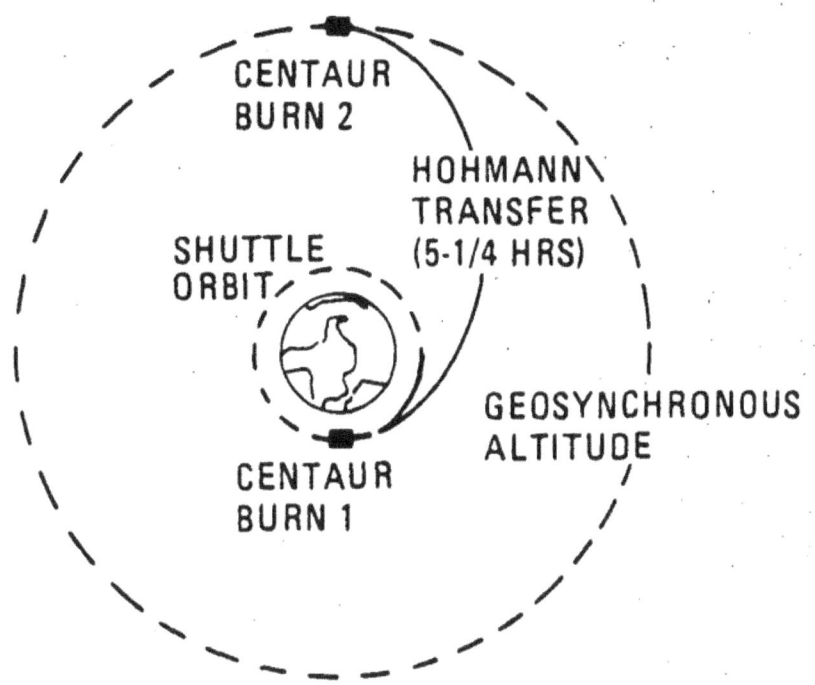

Figure 3. - Geosynchronous mission.

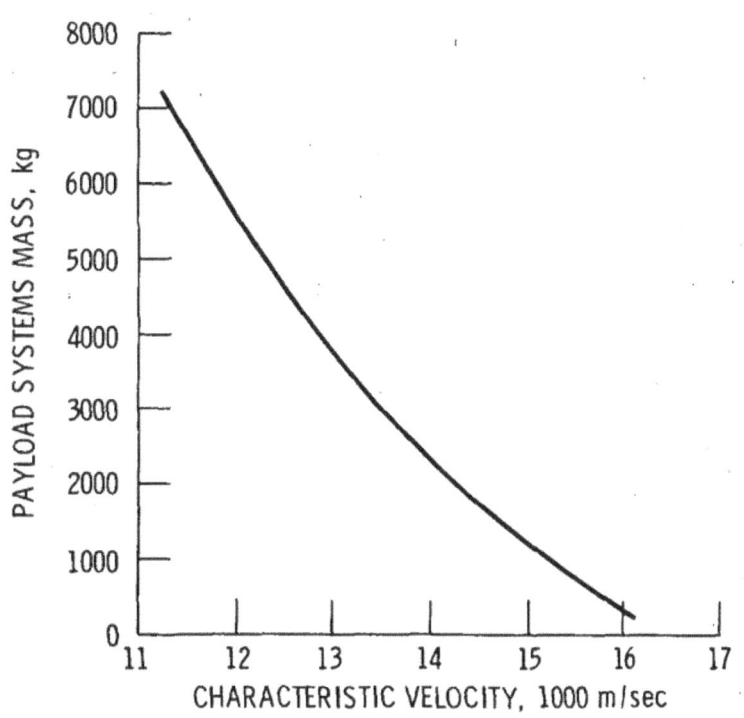

Figure 4. - Payload capability of Centaur G-Prime.

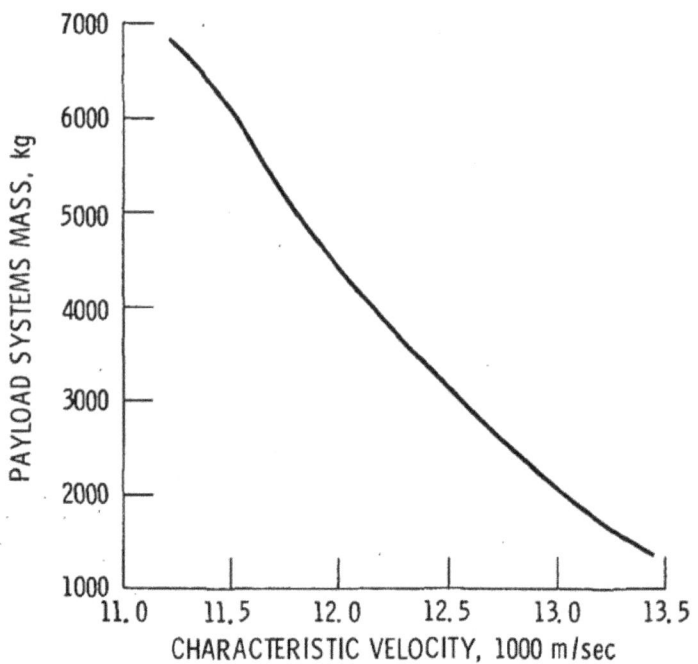

Figure 5. - Payload capability of Centaur G.

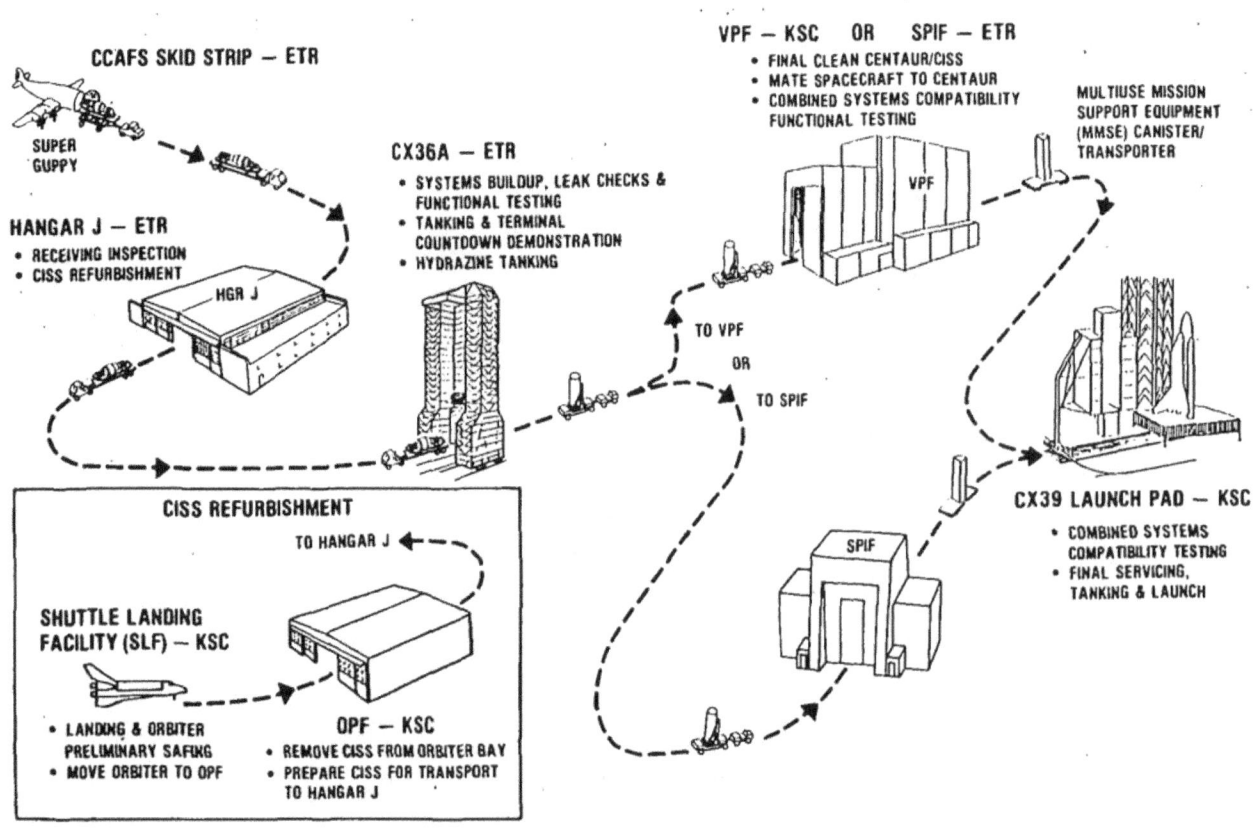

Figure 6. - Shuttle/Centaur launch operations.